Per Axel Rydberg's Botanical Collecting Trips to Western Nebraska in 1890 and 1891

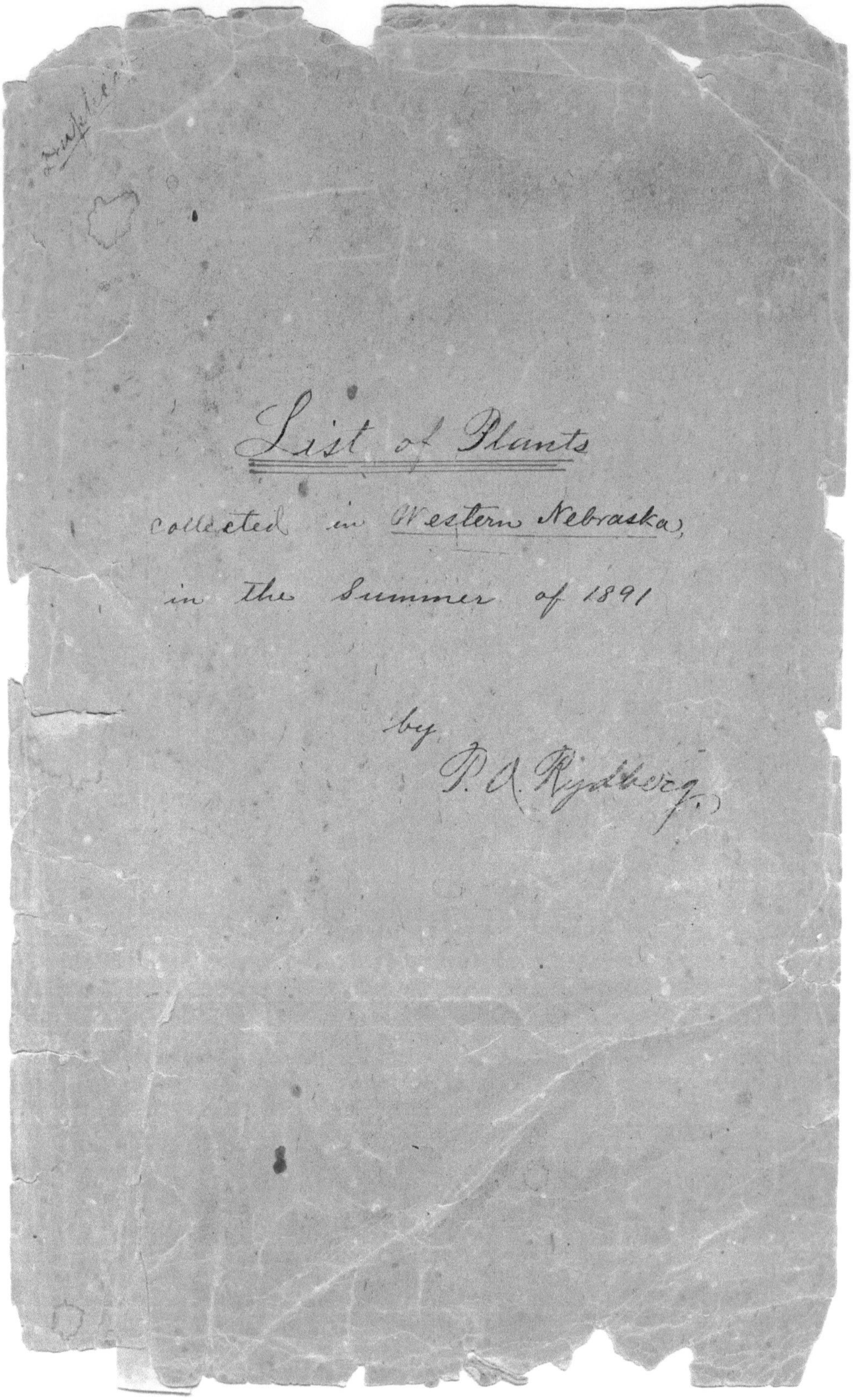
List of Plants

collected in Western Nebraska,

in the Summer of 1891

by

P. A. Rydberg.

Per Axel Rydberg's handwritten cover for his *List of Plants* collected in 1891.

Per Axel Rydberg's Botanical Collecting Trips to Western Nebraska in 1890 and 1891

His *Report* and *List* of Specimens for 1891, Annotated and with Notes about the 1890 Trip

Robert B. Kaul
The Bessey Herbarium (NEB)
University of Nebraska–Lincoln
Lincoln, NE 68588-0514

David M. Sutherland
Department of Biology
University of Nebraska at Omaha
Omaha, NE 68182-0040

Zea Books: Lincoln, Nebraska
2016

ISBN 978-1-60962-085-1 paperback

ISBN 978-1-60962-086-8 electronic book

Composed in Century Schoolbook, Elephant, and Segoe UI types by Paul Royster.

Zea Books are published by the
University of Nebraska–Lincoln Libraries.

Digital ebook edition (pdf) available at
http://digitalcommons.unl.edu

Print edition can be ordered from
http://www.lulu.com/spotlight/unllib

University of Nebraska Lincoln

Contents

Acknowledgments

We are grateful to Martha Kaul for her help with fieldwork and photography, and to Angie Fox for making the beautiful maps, figures 1 and 2. The University of Nebraska State Museum in Lincoln hosts the Bessey Herbarium, which dates from about 1874 and holds many of the Rydberg and Flodman specimens we have cited. The recent reorganization and nomenclatural updating of the collection by Linda Rader, Thomas Labedz, and student assistants have greatly assisted our efforts.

List of Figures and Maps

Introduction

In the summer of 1891, Per Axel Rydberg and his assistant, Julius Hjalmar Flodman, collected plants in western Nebraska for the United States Department of Agriculture. They collected many first-records for Nebraska as well as some that became *type* specimens of Rydberg's and other botanists' names. In the following autumn and winter, Rydberg made a detailed, typewritten, carbon copied 35-page ***Report*** and 37-page ***List*** (frontispiece) of specimens from that trip; one carbon copy is in the Bessey Herbarium (NEB) at the University of Nebraska–Lincoln. It is these documents that we present here, extensively annotated with our geographic clarifications, original and updated nomenclature, and citations of specimens in NEB and elsewhere. Rydberg, by himself, made a less extensive journey in western Nebraska a year earlier, 1890, and we have cited the specimens of the same species he collected that year in our annotated version of the *List*.

Sixty-five years ago, Walter B. Kiener (1894–1959), longtime and prolific botanical collector in Nebraska and Colorado (*Fuenning, 1960*), published the *Report* (but not the *List*) of the 1891 trip (*Kiener, 1951*). Kiener made geographical and historical footnotes but few botanical ones, and we have done so in both the *Report* and the *List*.

The 1890 and 1891 expeditions were among the first to do extensive collecting in most of the area traversed, which was then in early middle stages of agricultural development. The native flora was mostly intact but not well known, and nearly 80 years would pass before further intensive collecting was done there.

In September, 2012, we retraced the 1891 trip and photographed many of the sites that Rydberg and Flodman visited, and some of those photographs are given here. Rydberg and Flodman did not take photographs on that trip.

We compiled a list of all the specimens from the 1891 trip that are housed in the Bessey Herbarium, as well as the trip's specimens listed online (as of September 21, 2015) in the collections of the New York Botanical Garden (NY), the United States National Herbarium (US) in Washington, D.C., the Rocky Mountain Herbarium (RM) at the University of Wyoming, and a few other places.

The Collectors and Annotators

Per Axel Rydberg

The life of Per Axel Rydberg (1860–1931) (Fig. 3) is detailed in Rydberg's (1907) *Scandinavians who have contributed to the knowledge of the flora of North America*; Barnhart's (1931) *Per Axel Rydberg;* and Tiehm and Stafleu's (1990) *Per Axel Rydberg: A biography, bibliography, and list of his taxa.* Born in

Sweden on July 6, 1860, he emigrated to the United States in 1882, first to mining camps in upper Michigan and eventually to Wahoo, in Saunders County, Nebraska, where he taught mathematics from 1884 to 1893 at now long-defunct Luther Academy (Fig. 6), a Swedish-Lutheran school (Rydberg, 1907). In the summers of 1891-1893, 1895, and 1896 he was a field agent for the United States Department of Agriculture and collected plants in Nebraska, South Dakota, Montana, Idaho, Wyoming, and Colorado, and specimens from those and other trips are in the Bessey Herbarium.

Wahoo is 25 miles north of Lincoln (Fig. 1), the state capital and site of the University of Nebraska–Lincoln, which houses what is now known as the Charles E. Bessey Herbarium (NEB), founded about 1874. Rydberg enrolled at the university to study botany with Bessey, earning his first degree in 1891 and his M.S. in 1895 (Rydberg, 1895a,c). In 1898, he received his Ph.D. under Nathaniel Lord Britton at Columbia University in New York City. He became affiliated with the New York Botanical Garden and spent his remaining career there, eventually becoming its herbarium's director. He continued his extensive fieldwork in North America, but never again in Nebraska, although he visited the university herbarium there in 1926 and annotated some of his specimens. He published many botanical books and papers and named many new taxa, and all are listed by Tiehm & Stafleu (1990). Some of those names have survived to this day, and some have been revived from synonymy in the current nomenclatural upheaval brought upon by molecular and cladistic approaches. His taxonomic principles have taken their share of criticism, but he based them upon his extensive field and herbarium work and remained seemingly unfazed by criticism.

Rydberg's attention to detail and his stoicism and intractabiity (*cf.* Tiehm & Stafleu, 1991) are evident in the *Report*, but Flodman's thoughts are not in it. Rydberg seldom let primitive roads, balky horses, unreliable buggies, or rainy weather impede the pair's progress, but he occasionally complained about the accommodations and food at a few stops.

The Rydberg Specimens at the University of Nebraska–Lincoln

1883 Michigan (2 specimens)

1885, 1887, 1888 Nebraska (8)

1889 Nebraska (10), Michigan (11)

1890 Nebraska, mostly Sandhills and Panhandle (323)

1891 Western Nebraska (with Julius H. Flodman until August 14) (638)

1892 South Dakota: Black Hills (with A. E. Wagner until July 3) (771)

1893 Nebraska Sandhills (with N. P. Tulen until August 25) (893)

1894 Washington D.C. and vicinity (136)

1895 Nebraska (15), Colorado (13)

1895 Montana (22) and Idaho (3) (with Cornelius L. Shear)

1896 Montana (with Julius H. Flodman; most specimens are in only Flodman's name, and most were determined by Rydberg) (73)

1897 Montana, Wyoming, Idaho (mostly Yellowstone National Park, with Ernst A. Bessey, son of Charles E. Bessey) (1,005)

1900 Southern Colorado (with Frederick K. Vreeland) (454)

1929 Southern Kansas (with Ralph H. Imler) (248)

Julius H. Flodman

For most of the 1891 trip, Rydberg had the able assistance of his fellow teacher at Luther Academy, Julius Hjalmer Flodman (Fig. 4). Rydberg often flatteringly mentioned Flodman in the *Report* but seems to have treated him rather harshly on the expedition, as is occasionally evident in the *Report*. Flodman, who was not a botanist, collected with Rydberg on much of the 1891 trip, but his name is not on most of the specimens' labels. He collected a few specimens under his own name that year, but none of the specimens from 1891 at NEB have both collectors' names on the labels. By contrast, most labels on Flodman's specimens from their 1896 trip to Montana show Flodman as collector and Rydberg as identifier, not as co-collector.

Mr. Flodman was born in Sweden on September 23, 1859, and emigrated with his family to Polk Co., Nebraska, in 1872. After he graduated from Augustana College, Illinois, in 1890, he took a position at Luther Academy in Saunders County, there joining Rydberg, who had been there since 1884. He taught mathematics, civil government, reading, spelling, Latin, and Swedish Literature. In Flodman's early years at Luther, he took trips with Rydberg to western Nebraska in 1891 and Montana in 1896. On the Montana trip, Flodman collected and distributed specimens under his own name, and 73 of those are in the Bessey Herbarium. In his collection that year was a thistle that Rydberg later used as the type specimen of a new species, *Carduus flodmanii* Rydberg (holotype at NY, isotype at US, no voucher at NEB); the plant is now called *Cirsium flodmanii* (Rydberg) Arthur. Flodman spent the remainder of his career at Luther Academy. He died in 1949, age 90, and is buried in Sunrise Cemetery in Wahoo, Nebraska (Fig. 5).

John M. Holzinger

John M. Holzinger (1853–1929), botanist in the United States Department of Agriculture in Washington, D.C., from 1891 to 1893, reviewed Rydberg's specimens and commented upon their identifications. Those comments are listed together in Rydberg's *List* but are included individually with the relevant species in our version. Holzinger referred some specimens to such notable botanists of the time as Bebb, Vasey, and Millspaugh, as noted in the *Report* and *List*.

Holzinger later taught science and Latin for 35 years at Winona State Teachers College (now Winona State University) in Winona, Minnesota. He became a prominent bryologist and published about his extensive collections (*Winona Republican-Herald*, June 1, 1929). Rydberg and Flodman did not collect bryophytes on the 1891 trip.

The Specimens and Their Labels

Rydberg's Numbering System

On the 1891 trip, Rydberg numbered species, not specimens. Thus he gave all specimens of a given species the same number, regardless of their place and date of collection. For example, his specimens of *Ranunculus cymbalaria* from Banner County on July 17, 1891, have the same number, 5, as his specimens of that species from Scotts Bluff County on July 29 and August 1 of that year. Rydberg sometimes used the same numbers on other trips, e.g., his 232 is *Eleocharis engelmannii* in 1890, *Asclepias viridiflora* in 1891, and *Paspalum laeve* in 1929.

The labels of most of Rydberg's specimens from his 1890 trip to the Panhandle are unnumbered, but a few have other numbers that are someone else's, perhaps those of Herbert J. Webber, who kept a private herbarium in Lincoln that included many of Rydberg's specimens.

Spellings

We have retained Rydberg's botanical spellings, slightly modified his punctuation, and added, in brackets, current identifications and nomenclature, where needed. We have modernized his geographic spellings as shown below, with his spellings given first:

- Big Spring, Big Springs
- canon, canyon
- Chappel, Chappell
- Fort Kearny (with one "*e*") in Kearney County, near Kearney (the city in adjacent Buffalo County), both with two "*e*"s
- Kearney City, now Kearney (the city), in Buffalo, not Kearney County
- Kiwa, Kiowa
- Lodge Pole, Lodgepole
- McColligan Canon, McCuligan Canyon
- Pumpkin Seed Creek, Pumpkin Creek
- Pumpkin Seed Valley, Pumpkin Creek Valley

Scott's Bluff, now Scotts Bluff (the National Monument and the nearby bluff itself, both without the apostrophe); the present-day city of Scottsbluff (as one word) did not exist in 1891.

Wild Cat Mountains, Wildcat Hills

Clarification of Geographic Nomenclature

A few years after Rydberg and Flodman's trip in 1891, the boundaries of two counties on the itinerary were changed and two new county names were introduced. In 1891, Cheyenne and Deuel counties were very much larger than they are today. In 1908, each was divided into two counties, their smaller, southern parts retaining their original names and their larger, northern parts becoming Morrill and Garden counties, respectively. Thus, Rydberg could not cite Morrill or Garden counties, even though he and Flodman collected there, and we have noted them in the *List* and *Report*. Rush Creek, Magnesia Cliffs, and McCuligan Canyon are in modern Garden County (part of the larger Deuel County of the time), but Big Springs is in old and modern Deuel County. Pumpkin Creek and Lawrence Fork are mostly in today's Morrill County, but in Rydberg's time they were in the larger Cheyenne County of the time. All other counties mentioned in the *Report* and *List* were named as they were in 1891 and are today, but sometimes it is not clear which county is being mentioned.

Rydberg sometimes did not distinguish the North Platte River from the South Platte River, often calling both just "Platte River," and we have added that distinction. The two rivers rise in the Rocky Mountains of Colorado and unite near the city of North Platte in Lincoln County, Nebraska, to form today's Platte River (Fig. 1); Rydberg and Flodman collected near all three.

Kearney City, as it was known in 1891, is now Kearney, in Buffalo County (Fig.1) on the north side of the Platte River, but most of Rydberg's collections in the area are from adjacent Kearney *County*, across that river from Kearney City. He made most of his Kearney County collections on the outbound trip in June, before Flodman joined the expedition, but he made a few collections there on his homebound trip. On the outbound trip, he was hosted by Dr. Harry Hapeman of Minden (Fig. 1), in Kearney County, and himself a notable collector. Rydberg's only Buffalo County collections were in late August, when he collected in Kearney City and the nearby river-bottoms on his return trip to Lincoln.

Rydberg's references to "Scott's Bluff" are to both the county and the bluff itself (Fig. 19), which is now Scotts Bluff National Monument, but not to the city of that name (now spelled Scottsbluff), which was not founded and named until 1899 (*Perkey, 1995*). The names are now spelled without the apostrophe. Rydberg mentioned Gering (Fig. 1), a town established in 1887 (*Perkey, 1995*) and today the Scotts Bluff County seat and twin to larger Scottsbluff city, across the North Platte River.

Climate, Geology, and Soils

Rydberg and Flodman began collecting near the western limits of tallgrass prairies in central Nebraska and continued through mixed-grass prairies into the shortgrass prairies, sandhills, and pine-covered escarpments of what is today known as the Nebraska Panhandle, a name that Rydberg did not use and was perhaps not in general use then. The region is on the High Plains east of the Rocky Mountains, an area of strongly continental climate made semi-arid by the rain-shadow of the mountains. At 3,000 to 5,000 feet elevation, and with annual rainfall varying widely from 5 to 20 inches, the Panhandle is known for its droughts, deluges, and windiness. Winters sometimes have little snow but produce occasional heavy snows and blizzards. Summertime spells of regular afternoon showers and evening thunderstorms can alternate with short or extended droughts, even in the same season. The summer of 1891 was exceptionally wet, as Rydberg noted repeatedly in his *Report*; brief showers, all-day rains and violent storms were common. The flora was thus in excellent condition, but the travel conditions were made difficult by the mud and high waters. Normally dry depressions and streambeds were sometimes impassable or nearly so, and such high waters are perhaps why the collection has so few aquatic species. On July 30, 1891, Rydberg noted that "... from the showers we met with during our stay in western Nebraska, we should have come to the conclusion that western Nebraska was a land of rain rather than of drought." The pair collected in all but the worst weather, and they spent the evenings pressing the specimens and writing their notes.

Rydberg and Flodman did not collect on Sundays, participating instead in the religious and social activities noted in the *Report.* They scaled Courthouse Rock (Figs. 1, 18) in today's Morrill County (then part of the larger Cheyenne County of the time), on Saturday, the Fourth of July in 1891, where Flodman read aloud the Declaration of Independence from its summit, about 400 feet above the nearby North Platte River. This landmark of clay, sandstone, and volcanic ash, and capped with concretions, was mentioned by pioneers passing along the nearby Oregon, Mormon, and Pony Express trails.

The pair also collected on ponderosa pine-covered Scotts Bluff (Fig.19) and in the Wildcat Hills (Fig. 17), steep escarpments that rise abruptly from the surrounding plains (Fig. 2). Their exposed Oligocene, Miocene, and Pliocene strata date back to 30 million years (Prothero & Emry, 2004) and have been sculpted since then by wind, rain, and gravity. The collectors were also in the shallowly to steeply dissected gulches and canyons and on the flat tablelands and floodplains between, where the loamy to sandy or clayey soils vary from shallow to deep and are mostly stone-free.

The Flora

In Rydberg and Flodman's time, the area was in early middle stages of agricultural development, and the native flora was mostly intact. There were fewer cattle, crops, and weeds than today, but more horses. The bison had been eliminated by 1891, but pronghorn antelope and mule deer were common, and they are today. Now the non-arable parts are used mostly for cattle-grazing, and the level lands are occupied by pivot-irrigated fields of corn, soybeans, dry edible beans, and sugar-beets, while vast wheatfields are on the tablelands.

The western (Cordilleran) and eastern (Laurentide) Pleistocene glaciers did not cover the High Plains, and the climate warmed in post-glacial times. Evidence from fossil pollen shows that coniferous forest, including spruce, was at first interspersed with prairie but later gave way to treeless prairie. Exceptions are the bluffs and slopes, which are less prone to prairie fires, but there are no spruces in range today, although junipers and ponderosa pine are common there now.

The buttes and escarpments retain their late post-glacial woody vegetation of drought-tolerant plants, especially juniper (*Juniperus scopulorum*), ponderosa pine (*Pinus ponderosa* var. *scopulorum*), mountain-mahogany (*Cercocarpus montanus*), and skunkbrush (*Rhus aromatica*). The understory is a sparse community of deep-rooted, xerophytic and mostly perennial herbaceous plants, with *Yucca glauca* prominent. The level lands in the Panhandle were short- and mixed-grass prairies, now largely plowed under or heavily grazed, and the lands east of there were mixed-grass vegetation, also now much reduced by agriculture. Willows and other riparian trees lined the streams then and now, but elsewhere the woody vegetation was limited to shrubs.

The Itinerary

The locations indicated are shown on the maps in Figures 1 and 2 and photgraphs in Figures 7–19.

JUN 12 – Rydberg and Flodman arrived in Minden, KEARNEY CO. Overnight at Location 1.

JUN 13 — Collecting in KEARNEY CO. They got as far as the strip of sandhills south of the Platte River but stayed south of the river. Overnight at Location 1.

JUN 14 — Sunday (no botanizing) KEARNEY CO. Overnight at Location 1.

JUN 15 — Botanized in KEARNEY CO. and went

as far north as the islands in the Platte River at edge of BUFFALO CO. Overnight at Location 1.

JUN 16 — Pressed material and went west into Phelps Co. but did not collect there. Still based in KEARNEY CO. northwest of Axtell. Overnight at Location 2.

JUN 17 — Rain, no collecting, still based in KEARNEY CO. Overnight at Location 2.

JUN 18 — Collections made in PHELPS CO. Went as for north as the Platte River but did not cross it. Overnight at Location 2.

JUN 19 — Botanized in PHELPS CO. and took a train to Holdrege, PHELPS CO. Overnight at Location 3.

JUN 20 — Took a train to Bertrand, near border of GOSPER CO. Botanized inside GOSPER CO. Overnight at Location 4.

JUN 21 — Sunday was spent near Bertrand, no botanizing. Overnight at Location 4.

JUN 22 — Took a train to Curtis and botanized near there, FRONTIER CO. Stayed overnight in Curtis, Location 5.

JUN 23 — Took a train, commenting on the flora of Little Medicine Valley, but no collecting. Some collecting done at a stop at Lisbon, PERKINS CO. Stopped at Venango, PERKINS CO., and some collecting done there. Overnight in Venango, Location 6.

JUN 24 — Taken 35 mi north into DEUEL CO. Went to Rev Bergquist's north of Chappell. Overnight at Location 7.

JUN 25 — Collected S of Rev. Bergquist's, heading SE toward Big Springs. DEUEL CO. Overnight at Location 7.

JUN 26 — Went to McColligan [McCuligan] Canyon, north from Rev. Bergquist's. That was Deuel Co. then but is now GARDEN Co. Came back into DEUEL CO. Overnight at Location 7.

JUN 27 — Again collected southeast of Rev. Bergquist's, DEUEL CO. Overnight at Location 7.

JUN 28 — Sunday (no botanizing) DEUEL CO. Overnight at Location 7.

JUN 29 — Made arrangements for trip and got ready to collect in DEUEL CO. Overnight at Location 7.

JUN 30 — Started out but had horse problems and did not get underway, DEUEL CO. Overnight at Location 7.

JUL 1 — With replacement horse, started traveling north, entering what is now GARDEN CO. Overnight at Location 8.

JUL 2 — Collected near Rush Creek (Deuel Co. then, but now GARDEN CO.) Followed that creek north to the North Platte River and northwestward to Frank Anderson's house, perhaps near the border of what is now MORRILL CO. Overnight at Location 9.

JUL 3 — Collected in Cedar Creek and went to the mouth of Pumpkin Seed Creek in what is now MORRILL CO. Overnight at Location 10.

JUL 4 — Took Pumpkin Seed Creek Valley to Court House Rock, going through Redington and camping a little beyond, still in what is now MORRILL CO. Overnight at Location 11.

JUL 5 — Drove into BANNER CO. to a Swedish settlement in the southeast part, to E. Frostrom's place. Overnight at Location 12.

JUL 6 — It rained, but they got out to do some afternoon collecting nearby. BANNER CO. Overnight at Location 12.

JUL 7 — Moved from E. Frostrom's place to N. Frostrom's, to be nearer Lawrence Fork, BANNER CO. Overnight at Location 13.

JUL 8 — Drove to Lawrence Fork, collected there and on nearby hills, probably all in BANNER CO. Overnight at Location 14.

JUL 9 — Walked a few miles up the Fork, more collecting in BANNER, but probably not as far as Kimball Co. because Rydberg wrote that they went upstream only a mile or two. Overnight at Location 14.

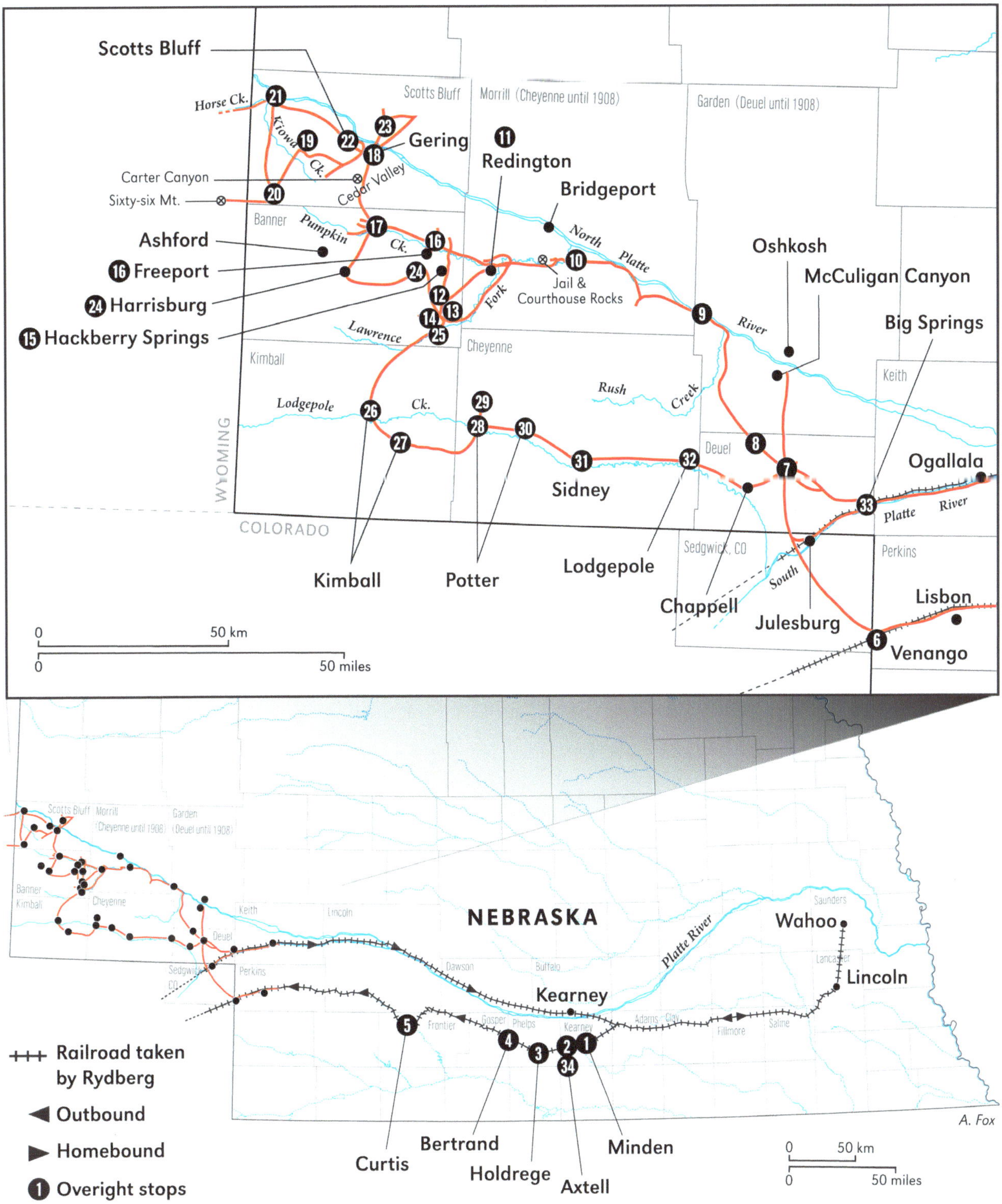

1. Maps of the 1891 trip. Numbered sites are overnight stops. Locations 1–6 were reached by train and the others by horse and buggy.

JUL 10 — Drove down the Lawrence Fork, back into what is now MORRILL CO. and all the way to juncture of Lawrence Fork with Pumpkin Seed Creek. Turned west and followed that Creek back to get into neighborhood of Hackberry Springs, BANNER CO. Overnight at Location 15.

JUL 11 — Visited Springs and nearby areas. Drove over divide and back to N. Frostrom's, BANNER CO. Overnight at Location 13.

JUL 12 — Sunday spent at N. Frostrom's, no botanizing, BANNER CO. Overnight at Location 13.

JUL 13 – Went back over the divide and into Pumpkin Seed Valley, crossed the valley to Freeport, camped in Freeport. BANNER CO. Overnight at Location 16.

JUL 14 — Collected in Wildcat Hills north of Pumpkin Seed Valley and went back to Freeport, BANNER CO. Overnight at Location 16.

JUL 15 — Moved camp from Freeport farther up the Valley to Ashford, which was apparently in the Valley at the base of the hills northeast of Harrisburg, BANNER CO. Overnight at Location 17.

JUL 16 — Collected in Wildcat Hills N of Ashford. Back to Ashford, BANNER CO. Overnight at Location 17.

JUL 17 — Went 5 or 6 miles farther up Pumpkin Seed Creek, then back to Ashford. BANNER CO. Overnight at Location 17.

JUL 18 — Took the stage road north over the Wildcat Hills and into SCOTTS BLUFF CO. Collections were made in Cedar Canyon. Camped just west of Gering, near the bluff itself.

JUL 18 — Sunday spent near Gering, no botanizing. SCOTTS BLUFF CO. Overnight at Location 18.

JUL 20 — Visited Scotts Bluff itself, w of Gering, following a canyon into area south of the Bluff. SCOTTS BLUFF CO. Overnight at Location 18.

JUL 21 — Followed the same canyon but went toward the North Platte River and collected in the canyon and in the North Platte bottomland. SCOTTS BLUFF CO. Overnight at Location 18.

JUL 22 — Botanized S of the Bluff and into what is now called Mitchell Pass, SCOTTS BLUFF CO. Overnight at Location 18.

JUL 23 — Botanized S of the Bluff, SCOTTS BLUFF CO. Overnight at Location 18.

JUL 24 — Botanized in badlands north of the Bluff, SCOTTS BLUFF CO. Overnight at Location 18.

JUL 25 — Drove southwestward into Carter Canyon and collected there, returning to Gering, SCOTTS BLUFF CO. Overnight at Location 18.

JUL 26 — Sunday spent near Gering, no botanizing, SCOTTS BLUFF CO. Overnight at Location 18.

JUL 27 — Drove west over Robidoux (Rydberg's "Roubideau") Pass and down into Kiowa Valley, SCOTTS BLUFF CO. Overnight at Location 19.

JUL 28 — Collected in and around Kiowa Valley, SCOTTS BLUFF CO. Overnight at Location 19.

JUL 29 — Went south into Pleasant Valley in southwestern SCOTTS BLUFF CO. and Sixtysix Mountain in adjacent GOSHEN CO., Wyoming. Overnight at Location 20.

JUL 30 — Stayed in this southwestern area in bad weather, SCOTTS BLUFF CO. Overnight at Location 20.

JUL 31 — Collected at Eagle's Nest in this area and took the road northward to the mouth of Horse Cr (where it joins North Platte River) in northwestern SCOTTS BLUFF CO. Overnight at Location 21.

AUG 1 — Visited a memorable ranch between Horse Creek and the North Platte River. Started out towards Gering and camped on the way, SCOTTS BLUFF CO. Overnight at Location 22.

AUG 2 — Sunday, rested in camp, no botanizing, SCOTTS BLUFF CO. Overnight at Location 22.

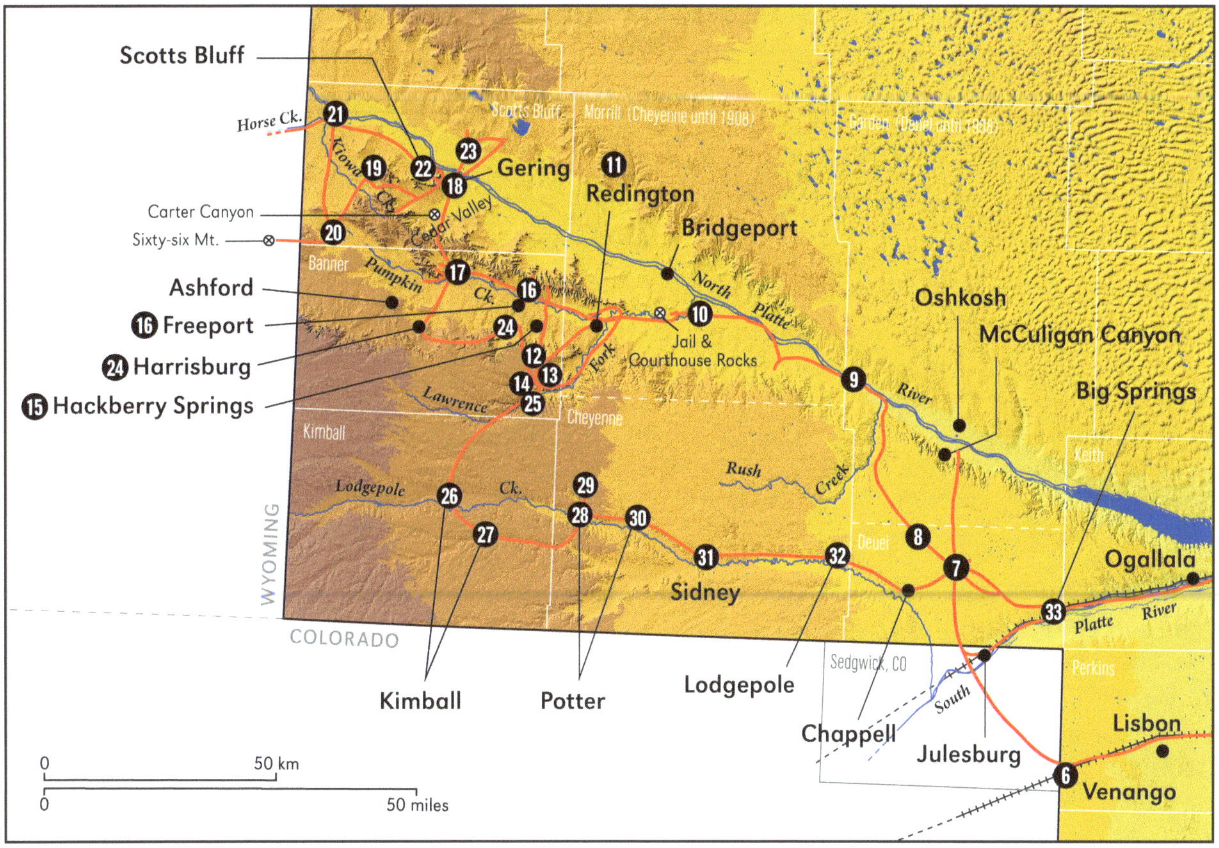

2. Topographic image of the Panhandle, with the trip's route overlaid. The broad valleys of the east-bound North and South Platte rivers have rugged terrain between them.

AUG 3 — Were about 5 miles from the Bluff in the Mitchell Valley, collected there and in bottomland on North Platte River, arriving back at Gering in afternoon, SCOTTS BLUFF CO. Overnight at Location 18.

AUG 4 — Crossed the North Platte River, turned west and went upstream 5 or 6 miles, turned S, and camped, SCOTTS BLUFF CO. Overnight at Location 23.

AUG 5 — Drove a distance eastward, then turned south back to the North Platte River and went back to Gering. Broke the wagon's double tree and made a temporary patch, SCOTTS BLUFF CO. Overnight at Location 18.

AUG 6 — Rydberg went back to the Bluff while Flodman got the double tree fixed (SCOTTS BLUFF CO.) They drove in afternoon back to Ashford, BANNER CO. Overnight at Location 17.

AUG 7 — Took road to Harrisburg, turned eastward and stopped and camped in a valley, BANNER CO. Overnight at Location 24.

AUG 8 — Botanized near camp and then in afternoon drove into Big Horn Valley and then ascended the table land between Pumpkin Seed Valley and Lawrence Fork, arriving back at E. Frostrom's, BANNER CO. Overnight at Location 12.

AUG 9 — Sunday spent at E. Frostrom's, no botany, BANNER CO. Overnight at Location 12.

AUG 10 — They drove across a divide and down into the Lawrence Fork, turning west up one of the branches, then turning south-southwestward and crossing a big draw (perhaps the branch now called Rocky Hollow), leaving BANNER CO., and entering KIMBALL CO. Overnight at Location 25.

AUG 11 — They collected in and near that draw, which Rydberg identified as a branch of the Lawrence Fork, and then drove on to Kimball, where they camped, KIMBALL CO. Overnight at Location 26.

AUG 12 — Botanized around Kimball, KIMBALL CO. Overnight at Location 26.

AUG 13 – They took the trunk and boxes to the railway station, sending material to Lincoln. Started toward Sidney, going southeast from town, and stoped at a ranch to camp, KIMBALL CO. Overnight at Location 27.

AUG 14 — Continued moving through KIMBALL CO., eventually entering CHEYENNE CO., where they turned N toward Potter. Flodman took the train to Lincoln that evening from Potter. Overnight at Location 28.

AUG 15 – Rydberg botanized near Potter and went to visit a friend 4 miles north of Potter, CHEYENNE CO. Overnight at Location 29.

AUG 16 — Rydberg spent Sunday with a friend, no botanizing, CHEYENNE CO. Overnight at Location 29.

AUG 17 — Collected in area north of Potter and started east, camped, CHEYENNE CO. Overnight at Location 30.

AUG 18 — Traveled to Sidney and just west of it, CHEYENNE CO. Overnight at Location 31.

AUG 19 — Collected near Sidney at the Lodge Pole Creek and on to Lodgepole town, arriving there late, CHEYENNE CO. Overnight at Location 32.

AUG 20 — Botanized near Lodgepole, CHEYENNE CO., and then moved on to Chappell in DEUEL CO. and all the way to Rev. Bergquist's place. Overnight at Location 7.

AUG 21 — Rested at Bergquist's, DEUEL CO. Overnight at Location 7.

AUG 22 — Went on family picnic back to McColligan (McCuligan) Canyon in today's GARDEN CO. Overnight at Location 7.

AUG 23 — Sunday, no botanizing, DEUEL CO. Overnight at Location 7.

AUG 24 — Botanized south of Rev. Bergquist's place and then pressed on toward, but not all the way to Julesburg. Started toward Big Springs, where he arrived late and took a hotel, DEUEL CO. Overnight at Location 33.

AUG 25 — Botanized along South Platte River before taking the road back to Rev. Bergquist's, DEUEL CO. Overnight at Location 7.

AUG 26 — Church festival and business in Chappell, DEUEL CO. Overnight at Location 7.

AUG 27 — Botanized around Rev. Bergquist's place, then went back to Chappell, DEUEL CO. Overnight at Location 7

AUG 28 – Rydberg went back to Julesburg, where he botanized a little on bottomland. Took the eastbound train bound for Kearney City. The night spent on the train, so no location on map.

AUG 29 — Arrived at Kearney in the morning and botanized along the Platte River (the border between BUFFALO and KEARNEY counties), but a change of plans then took him back to Axtell, in KEARNEY CO., where he stayed over with Rev. Thoren. Overnight at Location 33.

AUG 30 — Sunday spent with Rev. Thoren, no botanizing. KEARNEY CO. Overnight at Location 33.

AUG 31 – Rydberg took the train to Lincoln from Axtell.

Color plates

3. Per Axel Rydberg at about age 31, when he made the trip in 1891.

4. Julius Hjalmer Flodman, Rydberg's assistant on the 1891 Nebraska trip and again on the Montana trip in 1896.

5. Flodman's tombstone in Sunrise Cemetery, Wahoo, Nebraska.

6. Luther Academy in Wahoo, Nebraska, as shown on a postcard. No buildings remain today.

7. The South Platte River, shallow and sandy, near Julesburg, Colorado, where Rydberg collected *Cleome angustifolia*, *Bidens cernua*, and *Rayjacksonia annua*.

8. Pumpkin Creek, east of Redington.

9. Lawrence Fork of Pumpkin Creek, east of Redington.

10. Horse Creek in Goshen County, Wyoming, very near its entrance into Nebraska.

11. Shortgrass prairie between Sidney and Kimball.

12. Lodgepole Creek, in Cheyenne County.

13. McCuligan Canyon, in today's Garden County but the Deuel County of Rydberg's time.

14. Canyon near Bertrand (location 4) in Gosper County.

15. Carter Canyon, Scotts Bluff County.

16. Cedar Canyon, Scotts Bluff County.

17. Wildcat Hills, which Rydberg called Wild Cat Mountains. The trees are Rocky Mountain ponderosa pine, *Pinus ponderosa* var. *scopulorum*.

18. Courthouse and Jail Rocks, in Morrill County today but the Cheyenne County of Rydberg's time.

19. Scotts Bluff, Scotts Bluff County, with ponderosa pine. Rydberg and Flodman scaled the bluff and collected around it.

Rydberg's Report

Verbatim from his typescript, except for bracketed material, modernized spellings, and some modified punctuation. Most of the sites visited are shown in the maps (Figures 1 and 2).

REPORT
of a Botanical Exploration, made in Western Nebraska. In the summer of 1891, as a Special Agent of the Department of Agriculture

Hon. Dr. George Vasey,
Botanist, Dept. Ag.,
Washington, D. C.

Dear Sir:—

I have the honor to submit hereby my report of the work done by me as Special Agent of the Department of Agriculture in the summer of 1891.

Upon the recommendation of Dr. Charles E. Bessey, I was commissioned May 27, 1891, as a Special Agent of the Botanical Division of the Department of Agriculture, for the purpose of making a botanical exploration and of collecting specimens of plants of western Nebraska, under the direction of the Botanist of the Department, the time to be occupied being from the 12th day of June to the 12th day of Sept., 1891. By some cause, the commission was delayed and remained for some time in the post office at Lincoln and did not reach me until the 15th of July at Freeport, Banner Co., Nebraska. I had then been in the field over a month.

The day after the Commencement at the University of Nebraska, namely on June 12th, I began my journey westward, not waiting on my commission. The plans for the work had been laid out by Dr. Bessey and myself. I was going to stop over a day or two here and there on the road when passing through central and western Nebraska. The field proper was to be Deuel [including modern Garden], Cheyenne [including modern Morrill], Banner, and Scotts Bluff Counties, and, if possible, extend to Alliance in Box Butte Co. This plan was a little modified, however, as I saw that I would not have time to extend the trip even to the northern part of Scotts Bluff Co. On the return, I chose a different route, which took me through a part of Kimball Co., and through the Lodgepole Valley in Cheyenne and Deuel counties.

The determination of the plants collected and the preparation of the list with notes occupied my time between the first and the twelfth of September and, after that time, most Saturdays and other spare time from my work as teacher.

Having before only done the work of a student and amateur botanist—this being the first time I hold an appointment from the Government—I feel that my field-work, as well as my report, have many defects. I have tried my best. How I have succeeded can best be judged from the collection sent in, from my list of plants, and from the itinerary following below. My collection might have been larger as far as specimens are concerned, but as I was traveling for most of the time by team, 25 to 50 miles from any railroad, I could not very well bring with me more paper than I did. I was running five presses, and most the time they were overloaded with plants. The collection could also have contained a few more species. I forgot to collect specimens of some of the more common plants, as for instance of *Pinus ponderosa* var. *scopulorum*, *Fraxinus viridis* [F. pennsylvanica], *Ulmus americana*, *Juniperus virginiana*, etc. My list contains 479 species and varieties, not a very large number, but it must be taken in consideration that the flora of Nebraska is not very rich, and especially is this the case with the flora of the arid plains, hills and canyons of western Nebraska. My list contains about *two fifths* of all the *Anthophytes* reported for Neb. Of these, six are in Mr. Herbert Webber's CATALOGUE OF THE FLORA OF NEBRASKA without given localities, and 150 are not mentioned at all. Of these latter, 49 have been collected before in Nebraska, the most by me in 1890, some by Mr. A. T. [*sic*, Thomas A.] Williams, some by Rev. Bates of Valentine, and some by others. If my trip has not added much to the knowledge of the American flora in general, the list of Nebraska plants at the University has been increased by it by 10%, and if I count in the addition made by me last year in the same region, by about 13%. Of course, the list of new plants has been somewhat swelled by counting in all my varieties, some of which are mere forms.

In determining the plants, I have had access to the Herbarium of the University of Nebraska [international code NEB], which in many cases has been a good help. But as many of mine are new to Nebraska, much over 100 are not represented there by any specimens. Consequently, many could have been better determined, had I had access to a more complete collection.

My notes and remarks upon the plants collected are incorporated in my List of Plants; my observations of the country, soil, climate, and general flora, will be contained in the following.

ITINERARY.

June 12. — After having made two presses and packed my trunk with paper, I left Lincoln at 12 o'clock, noon, by the Chicago, Burlington and Quincy Railroad Main Line. At Fairmont, I was met by Mr. Julius H. Flodman, A.B., who made me company upon my trip for his own pleasure. To him I am very much indebted for the work he did as well as for his agreeable company. In fact, he did nearly as much work in the field as I, and to him is due the discovery of several of the rarer plants we found, as, for instance, *Stephanomeria minor* [S. runcinata], *Phlox caespitosa* var. *rigida* [P. hoodii], *Phlox muscoides*, and *Lesquerella engelmannii* [Physaria ovalifolia var. ovalifolia].

At 6 o'clock, we arrived in Minden, Kearney Co. One of my former pupils, Mr. Ed. Peterson, who lives near Minden, had promised to take us out in the country. I had written a card that we were to come on June 12. He had not been in town for several days, had not received my card, and, consequently, did not meet us. I met, however, a physician, Dr. Hapeman, who invited us to his home. Both he and his wife were very interested in botany. The evening was spent in looking over their collection of plants.

June 13. – After a good breakfast, the Doctor and Mr. Palmer, another amateur botanist of Minden, took us out on a botanical excursion to a dry creek northeast of Minden. Among other plants collected were *Callirhoe involucrata* and *alcaeoides*, *Equisetum laevigatum* and *variegatum* [E. laevigatum], two *Carices* [C. gravida, C. brevior], *Salix longifolia* var. (See No. 369 of my list.) [S. interior], *Chenopodium album* [C. berlandieri] and *leptophyllum* [C. pratericola], the latter seeming to be the worst weed of the two.

Maybe the most important grass in the meadow was a *Poa* [P. arida], which I regard as a variety of *Poa tenuifolia* (See No. 457 of my list.). Although the foliage is far from as luxurious as that of *Poa pratensis*, this bluegrass is easier made into hay, and may be improved by cultivation. On certain soils, I think it would be more profitable than the Kentucky bluegrass. I found it on low, even a little alkaline soil, at several places throughout western Nebraska, as for instance, on the Platte bottomland, as well in Kearney and Phelps counties as in Deuel and Cheyenne counties.

Among prairie grasses may be mentioned *Koeleria cristata* [K. macrantha] and *Eatonia obtusata* [Sphenopholis obtusata]. But perhaps the most important wild grass, especially for haymaking, is *Agropyrum glaucum* [Elymus smithii], which has received a various set of names, viz. slough grass, pond grass, wheatgrass, wild wheat, bluestem, bluegrass, Colorado bluegrass, grama grass, and even buffalo grass. Strangely, I did not hear at any place that it was called "quackgrass," the old name for *Agropyrum repens* [Elymus repens]. In Kearney Co., many think that it has come in since the land has been settled, but others say that it was found in sloughs and near ponds, hence the names "slough grass" and "pond grass," and from there it has spread over the prairies. Kearney Co. has also been a part of the "Great American Desert," which contained all of the territory west of the Misssouri River. The rain fell only in the spring, and then in great showers. The water could not penetrate the sod, but rushed directly into the rivers. The latter part of the summer was a season of great drought. This is still the case, to a certain extent, in the most western part of Nebraska. *Agropyrum glaucum* [Elymus smithii], which could not stand the drought as well as the buffalo grass (*Buchloë dactyloides*), the grama (*Bouteloua oligostachya* [B. gracilis]), and the bunch grass or niggerhead (*Carex* sp., No. 400 of my list) [C. filifolia, NEB 192032] was confined to the lowlands only. But since the country has been broken, the rain has been distributed throughout the seasons, and the prairie fires have been checked, and *Agropyrum glaucum* [Elymus smithii], has spread rapidly over the prairie. The buffalo grass has become scarce, and the other highland grasses have apparently disappeared.

The central and southern part of Kearney Co. is a high prairie plain as far as the eye can see, with a very rich loam, and is now one of the richest counties in Nebraska. But only 16 years ago, when the county was first settled, it suffered much from drought, was half a desert, covered mostly by the short highland grasses, but without any native trees. Now the county has a different appearance—fine farms surrounded by groves of cottonwood, boxelder, green ash, black walnut, and catalpa; fertile fields, green prairies, making good pasture, and hayland. A visible change has taken place even in the last four years, viz. since 1887, when I spent a summer in the county.

Just as we were ready to eat our lunch, Mr. Flodman came up to me with a handful of branches of a low bush with 3-foliate leaves, wanting to know the name. It was the poison ivy, *Rhus toxicodendron* [Toxicodendron radicans var. rydbergii]. Mr. Flodman had seen the poison ivy in Illinois, where it always is a climber. *Rhus toxicodendron* is in Nebraska a low shrub, except in the woody region of the very eastern part of the state. "Ignorance is a crime." Mr. Flodman (and also Dr. Hapeman received a dose) was severely punished for this crime. For a week his hands, face, and other parts of his body were in a very bad condition, all covered with blisters and swellings. Next Monday, one eye was completely closed and the other nearly so.

From Dry Creek, we drove north-eastward. The land became more hilly and more sandy. Among other plants, we collected *Penstemon grandiflorum* [P. grandiflorus] and *humilis* [P. albidus], *Lathyrus ornatus* [L. decaphyllus], *Asclepias arenaria*, etc. We came also a mile or two into the sand-hills proper, where we collected, among others, *Festuca tenella* [Vulpia octoflora], *Talinum teretifolium* [Phemeranthus parviflorus], *Psoralea campestris* [Pediomelum digitatum], *Aphyllon fasciculatum* [Orobanche fasciculata], etc., and arrived back at Minden late in the afternoon. After having taken an early supper, and putting the plants collected into press, Mr. Flodman and I took a 4-mile ride out into the country to spend the Sunday among friends. Late in the night we arrived at Mr. O. Peterson's.

June 14. — We attended the service at the Swedish Lutheran Church, visited in the afternoon Rev. Thoren and Mr. Otto Abrahmson, one of the first settlers of Kearney Co., and returned to Mr. Peterson's in the evening.

June 15. — Early in the morning, Mr. Peterson took us to Minden, where Dr. Hapeman and Mr. Palmer waited for us for another expedition to the Sand-hills and Platte River. We passed through the Sand-hills as fast as we could, that is, sometimes, step by step, picking up *Astragalus mollissimus* and *Evolvulus argenteus* [E. nuttallianus] on the way, intending to do the botanizing on the return.

Along the Platte, there are extensive natural haylands. Which the principal grasses are I cannot tell, as I do not know the grasses well enough to distinguish them by the leaves only, and very few were in bloom. I will enumerate a few I noticed: *Agropyrum glaucum* [Elymus smithii ?], *Poa tenuifolia* var. (No. 457) [P. arida], *Phalaris arundinacea*, *Calamagrostis canadensis* and *confinis* [C. stricta] (Nos. 430 and 441), *Panicum virgatum*, *scoparium* [P. leibergii?], and *dichotomum* var. *barbulatum* [P. acuminatum?], a *Muhlenbergia*, *Sporobolus cryptandrus*, *Agrostis alba* [A. stolonifera], *Phragmites communis* [P. australis], *Glyceria distans* [Puccinellia nuttalliana] and *nervata* [G. striata], etc. Among the grasses grew also *Triglochin maritima* and a narrow-leaved, long-petioled plantain near *Plantago eriopoda* (See my list, No. 304)

[P. eriopoda]; and in the water several species of *Scirpus.*

As we were going to visit some of the islands in the Platte, it began to rain. I had taken neither raincoat nor umbrella, as the day was clear when we started. I took one of the curtains from the topbuggy over my shoulders, and, followed by Dr. Hapeman and Mr. Palmer, crossed the railroad bridge. Mr. Flodman, who suffered very much from the poisoning, remained in the buggy, taking charge of one of the horses. On the railroad, we collected *Nasturtium sinuatum* [Rorippa sinuata], *Sinapis arvensis, Physalis virginiana* and *lanceolata* [P. hispida]. On the islands we found *Habenaria* [Platanthera praeclara, NY 24000, not on Rydberg's List, and no specimen in NEB], *Aspidium thelypteris* [Thelypteris palustris var. pubescens], *Equisetum limosum* [E. laevigatum], *Smilacina stellata* [Maianthemum stellatum], *Apios tuberosa* [A. americana], and *Amphicarpaea monoica* [A. bracteata].

On the islands and banks of the Platte River is found the whole of the native woody flora of the county (with the exception of such small bushes as *Rosa arkansana* and *Amorpha canescens*). The trees of the Platte have been of good size, but as the Platte bottom for many years furnished the fuel to the settlers of Kearney and Buffalo counties, no big trees are found except the planted cottonwoods around old Fort Kearney [*sic*, Fort Kearny]. The woody flora consists mostly of willows, *Salix amygdaloides, cordata* var. *vestita* [S. eriocephala], and several varieties of *longifolia* [S. interior]. More sparingly are found boxelder, green ash, red cedar, white elm, and cottonwood. Of the smaller woody plants, I saw *Celastrus scandens, Cornus stolonifera* [C. sericea], *Symphoricarpos occidentalis* and *vulgaris* [S. orbiculatus], and *Ampelopsis* [Parthenocissus vitacea], *Vitis riparia, Amorpha fruticosa, Rhus toxicodendron* [Toxicodendron radicans var. rydbergii], etc. When we returned, the rain had ceased, and the sun shone again; Mr. Palmer's driver had tied his horses and was gone down to the river; Mr. Flodman was sleeping in the buggy; the horse strolling around, enjoying the knee-deep grass.

As I have stated, the central and southern part of Kearney Co. is a high prairie, the northern part, except the Platte bottomland, is made up of a belt of sandhills, from 4 to 12 miles wide, unfit for cultivation, but partly used for pasturage for sheep. The sand is deep, and taking another route when returning, we had to drive and walk for several miles, botanizing at the same time. The flora is scanty, mostly consisting of grasses, among others, *Muhlenbergia pungens*, here called sand grass, *Stipa comata* and *viridula, Andropogon furcatus* [A. gerardii subsp. gerardii], *Sporobolus cryptandrus, Festuca tenella* [Vulpia octoflora], etc. Other plants seen were *Psoralea lanceolata* [Psoralidium lanceolatum], *Petalostemon villosus* [Dalea villosa], *Opuntia missouriensis*? [perhaps O. humifusa], *Lespedeza capitata*. We arrived very late in the night to Minden, where we were invited to Mr. Palmer's house.

June 16. — The day was cloudy and rainy. The forenoon was spent in pressing yesterday's collection, and changing paper for the plants collected on Saturday. A little before noon, we took the train to Axtell, where we met Rev. Thoren, who took us to Rev. Kinell's, 5 miles west-northwest of Axtell, and a few miles within Phelps Co. The roads were muddy, the grass wet, small showers the whole evening. No botanizing.

June 17. – Heavy rains all the day. We had to be confined to the house. No botanizing.

June 18. – Still raining. I changed papers in one press. The others I had sent by express from Axtell to Holdrege. A little before noon, it cleared up. Although the roads were in a miserable condition, I got a team in the afternoon and took a trip northward.

Phelps Co. resembles Kearney Co., the large part being a high prairie land, in the northern part traversed by the sand-hill belt, which, however, extends only through half the length of the county. Just before I reached the sand-hills, I collected *Stipa viridula, Psoralea tenuifolia* [*sic*, for *tenuiflora*] [Psoralidium tenuiflorum], (See my List, No. 48), *Polygala alba, Lippia cuneifolia*. In the sand-hills I collected *Mammillaria vivipara* [Coryphantha vivipara], *Psoralea campestris* [Pediomelum digitatum], and *Cnicus pitcheri* [Cirsium undulatum].

The Platte bottom was more sandy, and more alkaline than in Kearney Co. On the drier part grew extensive lumps of cactus, *Opuntia missouriensis* [O. polyacantha var. polyacantha] on the lower part, about the same plants as in Kearney Co., but the grass was much shorter. I saw also here and there patches of the salt grass, *Distichlis spicata* var. *stricta* [D. spicata], which is regarded as a sure sign of alkali. I also collected *Crepis runcinata* and a stinkhorn fungus, the smell of which was noticed by all as soon as I entered the house, when back.

The islands of the Platte were very few here, so also the trees. I saw some small willows, some cottonwoods, some boxelders, and some small plane-trees, *Platanus occidentalis*, very likely raised from seed, as the plane-tree grows only in the very southeastern corner of the state. I returned late at night.

June 19. – Cloudy; a few showers in the morning. I put my yesterday's collection in press and did a little botanizing around the place. Judging from the preceding day's trip, I concluded that it did not pay to stay any longer, and proposed to continue our journey. Rev. Kinell took us to Funk Station, where we took the train to Holdredge. Here we had to stop overnight, there being no connection between the trains.

June 20. – In the morning, we took the train to Bertrand, on the line between Phelps and Gosper counties. After having changed paper in the presses, we walked over to Mr. Swenson's, an old friend of Mr. Flodman's who lived half a mile from town. Mr. Swenson took us out to a canyon 6 miles west of Bertrand. Phelps Co. is very level, as previously stated; but as soon as Gosper Co. is entered, the land receives a different appearance— a rolling prairie traversed by canyons. One of these was the object of the expedition for today. However, we did not find very much of interest. *Yucca angustifolia* [Y. glauca], *Schrankia uncinata* [Mimosa quadrivalvis var. nuttallii], *Penstemon humilis* [P. albidus], and an *Astragalus*, which I have named *A. pubentissimus* (No. 74 of my List) [A. lotiflorus], were the most interesting. The grass on the hills was very short, and the soil seemed to be poor. The buffalo grass [Buchloë dactyloides] was more common. Of trees I did not see any except a few small willows, *Salix longifolia* [S. interior], a few stunted cottonwoods, and the western wild cherry, *Prunus demissa* [P. virginiana], all growing in the canyon.

We returned to Mr. Swenson's overnight.

June 21. – Being Sunday, we did not do any collecting, but attended church, made a visit to Rev. Sodergren, etc.

June 22. – In the morning we took the train to Curtis, Frontier Co. We traveled through Gosper Co. and the southeast corner of Lincoln Co. The land was very hilly and broken up by canyons. The soil was much poorer than in Phelps and Kearney counties. When we came down into the valley of Little Medicine Creek, the soil seemed to be a little better. In this valley is Curtis, the county seat of Frontier Co., situated. We arrived to this place at noon.

Immediately after dinner, we shouldered our collecting cases and set out for the creek. In the lake we saw a *Ranunculus* with dissected leaves, but I cannot tell which, as all the specimens were lost. On the railroad bank, we collected *Oenothera pinnatifida* [Oe. albicaulis], *Medicago sativa*, *Euphorbia serpyllifolia*, *Physalis virginiana* [P. heterophylla]. We then mounted one of the hills, where grew, among others, *Yucca angustifolia* [Y. glauca], *Psoralea esculenta* [Pediomelum esculentum] and *tenuifolia* [*sic*, *tenuiflora*] [Psoralidium tenuiflorum], *Rosa arkansana*. When going down the hill, we found a large number of the western cherry, *Prunus demissa* [P. virginiana]. We now entered the woods growing along the creek. The forest consists of cottonwood, boxelder, white and slippery elm, green ash, hackberry, wild plum, willows, *Salix amygdaloides* and *cordata* var. *vestita* [S. eriocephala]. Among smaller woody plants, we saw *Rosa fendleri* [R. woodsii] and *nutkana* [R. woodsii] (Nos. 101 and 100 of my list), *Cornus stolonifera* [C. sericea], *Symphoricarpos vulgaris* [S. orbiculatus], *Ribes gracile* [R. missouriense], *floridum* [R. americanum], and *aureum* [R. odoratum], *Vitis riparia*, and *Ampelopsis quinquefolia* [Parthenocissus vitacea, the only *Parthenocissus* in range].

A few days before, there had been a great hailstorm higher up the creek, which had caused a flood. The water had risen 6 or 7 feet over the banks and covered the grassland with mud. Against the railroad bridge there was heaped up a big pile of logs, some measuring up to 3 feet in diameter. This bridge led over a small tributary to Little Medicine Creek. In the creek itself, the water had been still higher. We followed the stream for a long distance in order to find a crossing. At last we found a place where a big log was placed across the creek, which log had been used as a walking bridge. The water reached about a foot over the log, and, furthermore, it had made another channel, about two rods wide and four feet deep. We took off our clothing. Mr. Flodman waded over the new channel; I handed over the clothes by means of a long pole, whereafter I also crossed, which was very hard, as the current was very strong. We then crossed the main channel by crawling over the log. On the bluffs on the other side of the creek we collected, among other plants, *Lupinus pusillus*, *Astragalus pubentisssimus* (No. 74 of my list) [A. lotiflorus], *missouriensis* (No. 73) [A. m. var. missouriensis], and *racemosus* (No. 70) [A. r. var. racemosus], *Oxytropis lambertii*, and, in a gorge, the red ash, *Fraxinus pubescens* [F. pennsylvanica], which is somewhat rare in Nebraska. We returned to Curtis about at sunset.

June 23. – In the forenoon we pressed our plants, wrote some letters, and took the train westward at noon.

The railroad followed the Little Medicine Valley for awhile. The valley consisted of rich bottomland, occasionally alkaline; *Distichlis* was growing here and there. The valley was surrounded by hills, dry and somewhat sandy. Further west the land became more sandy and more level, in some places very poor. The grass was generally very low. The buffalo grass and other highland grass became more common. Of the taller grasses, *Andropogon furcatus* [A. gerardii subsp. gerardii] and *Stipa comata* were very common. Not unusually, we saw big clumps of cactus, *Opuntia missouriensis* [Opuntia polyacantha?], and great patches of *Rumex venosus*.

At a little stopping place, whose name, if I remember right, was Lisbon, the wheels of the engine had run hot, and we had to stop for 40 minutes. We made use of the time in botanizing. We collected *Oenothera canescens*, *pinnatifida* [Oe. albicaulis], *coronopifolia*, and *hartwegii* var. *lavandulaefolia* [Oe. lavandulifolia], *Gaura coccinea*, *Argemone platyceras* [A. polyanthemos], *Sophora sericea* [S. nuttalliana], *Astragalus gracilis* and *microlobus* [A. gracilis], etc.

At 6 o'clock, P.M., we arrived at Venango, the last station in Nebraska on the Chicago, Burlington and Quincy Railroad. After supper, we went out for an hour or so. The most important of the collection made were *Echinospermum redowskii* var. *cupulatum* [Lappula redowskii var. cupulata], *Erigeron canus*, *Aster ericaefolius* [Chaetopappa ericoides], and *Plantago patagonica* var. *gnaphalodes* [P. patagonica var. patagonica], The latter is on dry prairies a widespread weed, which should not be left without consideration.

June 24. — In consideration of $6, the hotel keeper had promised to take us 35 miles northwest to the center of Deuel Co., Nebraska. Early in the morning, before sunrise, we packed my trunk, our valises, and presses on the wagon, and started off. On the road, we kept up a continuous conversation. Our driver had been living at Venango for five years and in western Nebraska for eight years. He gave us some information about the land, etc. There had been great drought the last two years, and the crop had been nearly none. This year there had been enough rain and he expected that the crop would be good, especially south of Venango, where the soil was better. Along the road where we were traveling, the soil was somewhat sandy. We passed a pretty good wheat field. The corn was late, which was the case all over in Nebraska, last spring. The fields were not so well tended to as they usually are in eastern Nebraska. This was the case nearly everywhere in the West. Undoubtedly the settlers had become a little indifferent when they had seen crop after crop fail.

The life on the prairie has changed very much in the last twenty years. The noble buffalo is exterminated from the soil of Nebraska. Our driver had not seen any since he came to western Nebraska. In the first years he saw "plenty" of antelopes, but in the last years he had seen very few. We did not see any during the whole summer.

The land became more and more hilly, and also more sandy, the nearer we came to South Platte River, which we crossed at Julesburg. Two miles north of Julesburg, we entered Nebraska again, being now in Deuel Co. Above the hills, the land became better. Nearly all the government land had been settled, but now most of the settlers had left their homesteads. Deuel Co. had not had a good crop for 5 years, and the homesteaders left as soon as they had proved up, and many long before. Last year many were still remaining in this part of the county. Some unscrupulous settlers and unscrupulous surveyors had made a scheme, an irrigation ditch. If I am not wrongly informed, the high tableland of Deuel Co. lies higher than the surrounding country. These men continued, it is said, to lay out irrigation ditches, even after they knew this was the case, and at last the headman of the enterprise disappeared. Disappointed, nearly all left that could leave. This, however, is not the case with a settlement of Swedes, of which the larger part still remain. Although originally, as a rule, poorer than their American, Irish, and German neighbors, have the Swedes been able to bear the hardships, by the untiring labor of Rev. F. W. Bergquist, who has appealed to the Swedish congregations of the country, and, as an answer, received and distributed among the needy carloads of grain and clothing.

We passed many vacuated cottages, many idle fields, and before we found anybody of whom we could ask for the road, we had driven 3 or 4 miles too far westward, which made our road 7 or 8 miles longer. We saw some leaf-cactus (*Opuntia*) with red flowers, which I took for a variety of *Opuntia missouriensis*, but which I afterwards found out to be *Opuntia rutila* [O. polyacantha var. polyacantha].

We arrived at Rev. Bergquist's in the afternoon, but as both Mr. Flodman and myself were tired from the long journey over the rough prairie, we did not do any collecting that day.

June 25. – Early in the morning we set out. Deuel Co. is a high tableland, with very few visible streams. The land is a little rolling, and the superfluous water, if there is any, is carried away through the "sand-draws." The sand-draw is a kind of dry river. That is, in the sand-draw is very seldom seen any water. It resembles a dried stream, with sandy or gravelly bottom. The sand is sometimes 15 to 20 feet deep or more. In the lower part of the sand, there is running a stream of water generally the year around. In one of the smaller sand-draws, I found a well, dug about 15 feet deep, that carried water in the month of August. Many springs and brooks of western Nebraska disappear in such sand-draws. At Potter, even the Lodgepole Creek sinks down in such a one to the depth of, it is said, 25 feet, and runs in that manner for several miles.

The sand-draws have a flora of their own. There are many plants that grow in the sand-draws which scarcely are found anywhere else. About 3/4 of the plants collected in Deuel Co. are collected in and near a sand-draw. It is strange that here, where the soil is very poor, clear sand or gravel should be a much richer flora than on the comparatively rich tableland. Undoubtedly, it depends solely upon the amount of moisture.

The field of our botanizing for the day was a sand-draw running south of Rev. Bergquist's, crossing Deuel Co. in a southeasterly direction and meeting the South Platte River a little above Big Springs. Before we reached this sand-draw, as we descended the hill, we collected a prostrate white *Petalostemon*, which I have given half a page in my list (See no. 58) [Dalea candida var. oligophylla]. We also collected *Astragalus sericoleucus* in fruit, *Paronychia jamesii*, *Malvastrum coccineum* [Sphaeralcea coccinea], *Linum rigidum* [L. compactum], *Erysimum asperum*, *Lesquerella ludoviciana* [Physaria ludoviciana]. *Krynitzkia crassisepala* [Cryptantha minima] and *jamesii* [Cryptantha cinerea var. jamesii] were very abundant. In the sand draw, we found *Oenothera albicaulis* var. (No. 112) and *serrulatus sic*, *Asclepias speciosa*, *Gilia linearis* [Collomia linearis] and *gracilis* [Microsteris gracilis var. humilior], *Euphorbia petaloidea* [E. missurica var. petaloidea], *Chrysopsis villosa* [Heterotheca villosa], *Polanisia trachysperma* [P. dodecandra subsp. trachysperma], and *Argemone platyceras* [A. polyanthemos], the last three not in bloom yet. On the prairie near the sand-draw, we found *Oenothera albicaulis* [Oe. nuttallii] and *coronopifolia* and *Potentilla pensylvanica*; and on a hill near the same, *Sophora sericea* [S. nuttalliana], *Lupinus pusillus*, *Psoralea hypogaea* [Pediomelum hypogaeum], *Opuntia missouriensis* [O. polyacantha?] and *gracilis* [O. fragilis?], *Cnicus undulatus* [Cirisium undulatum], etc. When returning we crossed the prairie and, among others, we collected *Astragalus caryocarpus* [A. crassicarpus] in fruit, *Oenothera cane-*

scens, *Thelesperma filifolium*, *Allium reticulatum* [A. drummondii], and *Castileia* [*sic*] *sessiliflora*. In an old field we found *Astragalus gracilis* and *microlobus* [A. gracilis]. In the evening we tended to the presses.

June 26. — I had written to Rev. Bergquist to buy or hire a team for the summer. He had just returned from a journey and had not had time after his return to do it. He, therefore, harnessed his own horses and gave us a 12-mile ride to a canyon, running through the bluffs of North Platte River. The canyon has received its name from a certain Mr. McCuligan, a noted horse-thief and robber, who for some time had his headquarter in the same. The canyon was about 200 feet deep. In the bottom, there are a few small springs, which give rise to a small brook, which is lost in the sand-draw below. Here we collected *Astragalus adsurgens* [A. laxmannii], *Oryzopsis cuspidata* [Stipa hymenoides], *Stipa viridula*, *Euphorbia petaloidea* [E. missurica var. petaloidea], *Croton texensis*, etc. On the hills surrounding the canyon, we found *Polygala alba*, *Astragalus caespitosus* [A. spatulatus], *Thelesperma gracile* [T. megapotamicum], *Actinella acaulis* [Tetraneuris acaulis], *Krynitzkia glomerata* [Cryptantha thyrsiflora] and *sericea* [Cryptantha cana] (No. 255 of my list), *Eriogonum alatum* and *flavum*, *Euphorbia montana* [E. robusta], *Stipa comata*, etc.

In McCuligan and other canyons, and on the surrounding hills, there has been a certain amount of timber, but as the remainder of Deuel Co. is altogether without any trees, these canyons have furnished the settlers with the larger part of their fuel and there is therefore now very little left. Of the Rocky Mountain yellow pine, I saw only stumps and dry limbs and one or two scragly bushes. The red cedar was more common, but there were no trees of any larger size left. In the canyon there grew some hardwood, as green ash, hackberry, white elm, boxelder, and cottonwood, but the trees were of small size. The larger part of the woody flora consisted, however, of the western cherry, *Prunus demissa* [P. virginiana] and of *Symphoricarpos*. The wild grapes, *Vitis riparia*, and the Virginia creeper, *Ampelopsis quinquefolia* [Parthenocissus vitacea, the only *Parthenocissus* in range], covered the bushes and the sides of the canyon. In the sand-draw we found also the sand cherry, *Prunus pumila* [P. pumila var. besseyi]. When driving home we collected *Petalostemon gracile* [Dalea candida var. oligophylla] in a wet meadow.

June 27. – We intended to spend the day in botanizing in the lower part of the same sand-draw we visited on June 25. In order to make the way shorter, we crossed the high prairie land.

The larger part of Deuel Co. south of North Platte River is a tableland covered with short grass. The most common grasses are buffalo grass (*Buchloe dactyloides*), the grama grass (*Bouteloua oligostachya* [B. gracilis]), and the bunchgrass, or wire-grass (*Carex* sp., No. 400 of my list) [C. filifolia]. The common names of these are badly mixed together by the settlers, and it is nearly impossible to know which is referred to, when, for instance, buffalo grass is mentioned. This name I have heard for all the three grasses mentioned, for *Bouteloua hirsuta* and *Munroa squarrosa*, occasionally for any short grass, as, for instance, *Muhlenbergia pungens*, *Festuca tenella* [Vulpia octoflora], *Sporobolus cuspidatus* [Muhlenbergia cuspidata] and *depauperatus* [Muhlenbergia richardsonis] and *Sc[h]edon[n]ardus texanus* [S. paniculatus]; and still more strangely, even for some taller grasses, as, for instance, *Agropyrum glaucum* [Elymus smithii?] and *Andropogon hallii* [A. gerardii subsp. hallii].

The grama grass seems to be the most important grass of the tablelands of western Nebraska. It is regarded as the best pasture grass and is preferred before any other by the animals. Our horses always selected it and grew fat, although we sometimes drove 20 to 25 miles a day over the prairies and never gave them as much as a handful of oats, corn, or any other feed. Whenever it grows a little taller, it gives an excellent hay. It never grows on really wet grounds, but on land that had plenty rain and good drainage, it reached a height of over two feet. Under irrigation, where the amount of moisture could be regulated, I think it will make a very excellent haygrass.

The buffalo grass [Buchloë dactyloides] is of less value, partly because it is not as common and partly because it is so short that it is even of little value as a pasture grass.

The bunchgrass is regarded as a valuable pasture grass, especially does it play an important part in the winter pasture. In the dry autumn, this as well as the two preceding become "self-cured," and, it is claimed, makes a very good pasture throughout the winter. The *Carices* are generally regarded as worthless, but this is, without any doubt, one of the exceptions. Many herders do not give their cattle any hay throughout the winter, except on stormy days and when the snow is too deep for the animals to scratch away with their hooves, which very seldom is the case. One sheep rancher of Deuel Co. used to drive a big snow-plow over the prairies and let his sheep graze in the furrow.

We struck a small branch of the sand-draw, which we followed. In this was found *Gilia linearis* [Collomia linearis] and *gracilis* [Microsteris gracilis var. humilior], *Hedeoma hispida, Eliocharis* [*sic*] *acicularis*, etc. On the prairie, near this draw, we collected *Oenothera albicaulis* [Oe. nuttallii], *serrulata, pinnatifida* [Oe. albicaulis], *sinuata* [Oe. laciniata], and *canescens, Gaura coccinea* var. (No. 119 of my list), *Koeleria cristata* var *gracilis* [K. macrantha], *Psoralea tenuiflora* [Psoralidium tenuiflorum] (See No. 48 of my list.). Further down, in the main sand-draw, we found *Pectis angustifolia*, a *Petalostemon*, which I have named *P. tenuifolius* Gray (See No. 59 of my list.) [Dalea purpurea var. purpurea], although I scarcely think it is specifically distinct from *P. violaceum*. On an alkaline spot near the draw we found *Monolepis chenopodioides* [M. nuttalliana], and, in a spring, *Limosella aquatica*.

When we went home, we struck several lumps of a red-flowered leaf-cactus, *Opuntia*. I had seen it before but always regarded it as a variety of *O. missouriensis*. I found, however, that the joints are much smaller than in *O. missouriensis*, and the spines different. I now regard it as *O. rutila* [O. polyacantha var. polyacantha]. I also found a yellow-flowered form of the same (No. 125b of my list), growing together with it. The cactuses of Deuel Co. are *O. missouriensis* (yellow), *rutila* (red, or yellow) [O. polyacantha var. polyacantha], and *gracilis* (yellow) [O. fragilis?] and *Mammillaria vivipara* (red) [Coryphantha vivipara]. A boy told me that there was also growing a yellow-flowered ball-cactus, (± *Mammillaria*), undoubtedly *M. missouriensis* [Coryphantha missouriensis], which has been collected at Valentine, Neb., by Rev. Bates. *Opuntia rafinesquii* Eng. [?] has also been collected in the state, but I have not seen it growing.

On the hills we also found *Erigeron pumilus*, a variety of *Penstemon*, which I have named *P. fremontii* T. & G. var. (No. 275 of my list) [P. albidus], *Aphyllon fasciculatum* [Orobanche fasciculata], and, on the prairie, *Draba caroliniana* var. *micrantha* [D. reptans], *Alopecurus geniculatus* [A. carolinianus], and *Festuca tenella* [Vulpia octoflora].

June 28. — Being Sunday, we did not do any botanizing.

June 29. — I made two more presses in the morning. We made arrangement to hire a team of horses for the summer of one of the neighbors, Mr. U. Jackson. I therefore drove down to Chappell, the county seat of Deuel Co., in order to buy provisions and to get my tent, which had been sent by freight to that place. The remainder of the day we were occupied in loading our wagon.

June 30. – In the morning we started. One of the horses was balky. Mr. Jackson had said that it was likely that it would be hard to get started, as it had been in pasture for a time. As soon as we had the horses hitched before the buggy, it turned squarely, putting its head over the back of the other horse. We could not make it go. Mr. Flodman tied a rope to the bridle and led it for awhile, running by its side. He then slacked off and mounted the moving wagon, with the rope still in his hand. The horse soon stopped again, and turned. So we kept on for a mile and a half, the horse turning a dozen times. We were then tired and returned, principally in the same manner. Rev. Bergquist was sitting in his cart, ready to go to Chappell. I left the horses in Mr. Flodman's care and mounted the cart, going along to town, to see if I could get another horse there. At last I found that Mr. Hamilton, a cashier of the Deuel Co. Bank, had a horse in his care, which I could hire, and so I did, bringing the horse with me home. I put it out on a rope to grass on the prairie. After supper, the hired boy came in and said that there was a horse in the barley field, and, surely, there our horse was. It had broken the halter and crawled under the one wire that served as a gate. When we tried to catch it, it ran towards the pasture where the other horses were. When they saw it, they began to run, and the horse followed on its side of the fence. At the end of the field, it broke through the wire fence and ran over the prairie towards Chappell.

July 1. – In the morning, Rev. Bergquist and I went to Chappell to find out if the horse was there and if it was hurt. They had seen it near the pasture where it used to be but could not tell if or how severely it was hurt. It had, however, been able to run the 9 miles from Rev. Berquist's to Chappell, and was, very likely, not severely damaged. At Gering, Scotts Bluff Co,, I received the notice that it had become stifled, and on my return I had to pay $30 for doctoring and damages. Mr. Hamilton offered me an old pony instead, which offer I accepted, and returned home. The old pony did scarcely look as if able to make such a journey as we had planned. It was nearly toothless, but still kept well, yea, grew fatter, on nothing but grass and served to full satisfaction.

In the afternoon we started anew and arrived in the evening to Mr. Nordell's, 12 miles north-northwest from Rev. Bergquist's. Mr. Nordell had not had a crop for four years, but he had been prudent enough to buy some cattle when he settled, and he and his family containing five children had lived during the hard years, principally by selling butter from a score or so of milk cows.

July 2. – We intended to follow Rush Creek down to North Platte River. Before we reached the bluffs, we crossed an old field overgrown with *Solanum rostratum*, *Chenopodium leptophyllum* var. *oblongifolium* [C. desiccatum], and *Amarantus* [*sic*] *blitoides* [A. albus]. The road was running down through a gorge from the tableland to the bottomland of Rush Creek. In this gorge we collected *Phlox douglasii* var. *longifolia* [P. andicola], and on the hills on the side thereof *Oenothera hartwegii* var. *lavandulaefolia* [Oe. lavandulifolia], *Krynitzkia glomerata* [Cryptantha thyrsiflora], and *Oxytropis lambertii* var. *sericea* [O. sericea]. The bottomland was used as pastures. We found big herds of cattle. The soil was sandy. Still, the grass did not seem to be bad, although it was well kept down by the cattle. What grasses grew there, I could not tell. I shall enumerate a few I noticed: *Bouteloua oligostachya* [B. gracilis], *Eatonia obtusata* [Sphenopholis obtusata], a *Muhlenbergia*, *Sporobolus airoides*, *Oryzopsis cuspidata* [Stipa hymenoides], *Munroa squarrosa*, *Elymus sitanion* [E. elymoides subsp. brevifolius], *Andropogon scoparius* [Schizachyrium scoparium] and *furcatus* [A. gerardii subsp. gerardii], and *Agropyrum glaucum* [Elymus smithii?]. Around the road there grew a great amount of weeds, as, for instance, *Cleome integrifolia* [C. serrulata], *Polanisia trachysperma* [P. dodecandra subsp. trachysperma], *Symphoricarpos occidentalis*, *Monarda fistulosa* [M. f. var. menthifolia], *Argemone platyceras* [A. polyanthemos], *Oenothera serrulata*, *Croton texensis*, and several species of *Helianthus*. We picked up a few plants on the road. We did not stop till in the afternoon, partly because we did not dare to leave our horses among the cattle, partly because there was not much to collect in the pastures. As soon as we could find a suitable place for the horses, we pitched them out and took our dinner, whereafter we spent a few hours in botanizing. In the valley, we collected *Thermopsis rhombifolia*, *Oenothera pinnatifida* [Oe. albicaulis] and *albicaulis* [nuttallii], *Physalis* sp. (No. 273) [P. hispida], *Monarda citriodora* [M. pectinata], *Abronia fragrans*, etc. On the surrounding hills we found *Erysimum asperum*, *Psoralea tenuiflora* [Psoralidium tenuiflorum], *Oxytropis lambertii* var. (No. 82c of my list) [O. lambertii], *Thelesperma filifolium*, *Actinella acaulis* [Tetraneuris acaulis], *Aphyllon fasciculatum* [Orobanche fasciculata], *Eriogonum flavum*, *Bouteloua racemosa*? [Bouteloua curtipendula], etc. After having put the plants collected into press, we started again and drove down the valley till we reached the North Platte River. Here we turned towards northwest, and followed the river upwards for 6 or 7 miles. We intended to stop at the first place we could get water, but we found that the road ran through one great pasture after the other, did not dare to picket out the horses in the pastures, and drove on until it was dark. We then came to Mr. Frank Anderson's house. Here we found water and a place where we could picket out the team. Mr. Anderson invited us into his house, where we stayed overnight.

July 3. – In the forenoon we traversed several pastures. The soil was mostly more or less sandy. The land seemed to be of very little value as farming land. The cattle ranchers, who have been obliged to vacate the land nearly all over western Nebraska, were here still holding the field. There was very little of interest in the way of plants, and we drove on in order to come to better fields as soon as possible. A few miles east of the mouth of Cedar Creek, we left the trail and followed a road which we supposed should lead into the valley of the creek, where we intended to spend some time in botanizing. When we came near the creek, we saw that it was flowing in a deep canyon. We were afraid that if we drove too far up we could not cross the same. The land also became more and more sandy. We followed the creek down to the next farmhouse. Here they had made use of the water of the creek and irrigated their small fields, but the crop did not seem to be very promising. The banks of the creek were very steep, and it was difficult for the horses to mount the sand on the other side. Here *Cycloloma platyphyllum* [C. atriplicifolium] was very abundant and also *Chenopodium leptophyllum* [C. pratericola]. Having taken a 3/4-hour rest and eaten our dinner, we continued our journey through sandhill after sandhill. At last we struck a road leading north-northwest. This we followed until we reached the California Trail again on the bottomland of the Platte. Here were excellent haylands for many miles. Most of the grasses were not in bloom yet. I collected, among others, *Oryzopsis cuspidata* [Stipa hymenoides], *Sporobolus airoides* and *asperifolius* [Muhlenbergia asperifolia], *Agrostis alba* [A. stolonifera], *Calamagrostis stricta*, *Schedon[n]ardus texensis* [S. paniculatus], *Distichlis spicata* var. *stricta* [D. spicata], etc. I also gathered *Juncus filiformis* [J. arcticus var. balticus] and *bufonius*, *Senecio aureus* var. (No. 211 of my list) [S. plattensis, Packera plattensis], *Crepis runcinata*, and *Dodecantheon* [*sic*] *meadea* [*sic*] [Dodecatheon pulchellum; the similar *D. meadia* does not occur in Nebraska].

When we came nearer to the mouth of the Pumpkin Creek, we found the land more settled. We saw several farms and ranches where the people were busy in preparing for the celebration of the Fourth of July. The most interesting we saw in the way of plants were some specimens of *Gaura parviflora*, 6 to 7 feet high.

In the evening there rose a black cloud very suddenly. We drove to a house in order to be in the neighborhood of water, picketed out our horses, and were just ready to pitch our tent when the shower came. We threw the tent over the buggy, tied an oilcloth thereover, and ran into the house. When the shower was over, we pitched the tent and carried in our things. It was raining and blowing hard nearly the whole night, and many times we feared the whole tent would blow away. The ground and the air in the tent were damp and chilly. It was the hardest night we had during the trip.

July 4. – In the morning, the wind and rain had ceased, the sky was clear. In the west was seen the renowned Courthouse Rock, which in the rays of the rising sun looked like a real court house with a cupola. Courthouse Rock is an isolated butte, over 300 feet high over the valley. It is situated a mile north of Pumpkin Creek and about 6 or 8 miles from its mouth. It has for ages been a landmark, and has received its name from the resemblance it bears to a courthouse, when seen from the east or from the west. Seen from the south, it resembles more the wall of an old ruin. On the east side the hills gradually rise up to about two fifths of the butte. This is here scarcely 50 feet wide. From here the west side is perpendicular up to the top or even a little overhanging. On the eastern side the rock can be mounted, but the path is winding and very steep; in two places steps have been hewn out in the rock; in one place, for a height of 9 or 10 feet, two rows of holes to put the hands and feet in, cut out in the perpendicular rock, furnishes the only possible way to reach the top. The rock of which the butte consists is nearly white, very soft (probably some kind of magnesia), and can be cut by a knife, but seems to withstand the action of wind and rain very well. A little southeast of the Courthouse Rock proper, there is another, smaller butte [Jail Rock], about 75 feet lower. This looks more like a big boulder, a broken pillar, or one of the flat-topped turrets on the walls of the ancient castles. No human foot has ever trod the top of it, and probably none will.

We had only 7 or 8 miles to the Courthouse Rock, but we had to drive several miles forth and back, because we could not find a bridge or a wade over Pumpkin Creek, so we did not reach it till towards noon. After having taken our dinner, we mounted the "Rock." Mr. Flodman, who used to study history on leisure hours, had taken Barnes' *U. S. History* with him and read the "Declaration of Independence" from the top of the mountain. In this way did we celebrate our Fourth of July.

On going down, we botanized, We had here an altogether different flora, which we found repeated on nearly all the mountains of Banner and Scotts Bluff counties. The most remarkable plants we found were *Arenaria franklinii* var. (No. 31 of my list) [Eremogone hookeri], *Astragalus caespitosus* [A. spatulatus], *triphyllus* [A. hyalinus], and *kentrophyta, Musenium* [*sic*] *tenuifolium, Hedeoma drummondii, Cycloloma platyphyllum* [C. atriplicifolium], *Eriogonum cernuum, flavum,* and *multiceps* [E. pauciflorum var. gnaphalodes], and *Rumex venosus.*

In the afternoon, we drove about 12 miles westward, passed through Redington, a little town of a couple dozen houses, about two-thirds of which were sod. Here the people of the neighborhood had gathered to celebrate the Fourth. We did not like to stop there, but drove a few miles more, before we put up our tent.

July 5. — Being Sunday, we did not do any botanizing but drove about 10 miles in the forenoon in order to reach the Swedish settlement in Banner Co. Rev. [Frank] Bergquist was that day visiting a small congregation there under his charge. He had promised to take with him a few things which we could not take, when we left his house. The settlement is in the high tableland between the Pumpkin Valley and the Lawrence Fork. The hills leading up to the divide are very steep, and we had to let the horses stop twice to rest, stones being put behind the wheels, before the top was reached. At noon, we arrived at Mr. Emil Frostrom's. Rev. Bergquist had already left the settlement and we did not meet.

July 6. — Heavy showers of rain in the forenoon, but in the afternoon it cleared up a little. Although the grass was very wet, we concluded to go out a while. We went to a canyon, about 200 feet deep, running from the valley into the tableland. On the tableland, just before we reached the canyon, we found *Cnicus undulatus* [Cirsium undulatum] and *ochrocentrus* [Cirsium ochrocentrum], the first in bloom but the latter just budding, and also a mountain mahogany, *Cercocarpus parvifolius* [C. montanus], which has not been reported for Nebraska before in 1890, when it was collected by me [NEB 175023, -4, -7]. In the canyon, among the cliffs, and on the sides of the hills, we found a number of plants, a partial list of which is the following: *Arenaria franklinii* var. *minor* [Eremogone hookeri], *Psoralea lanceolata* [Psoralidium lanceolatum], *Astragalus triphyllus* [A. hyalinus], *Oxytropis lambertii* var. (No. 82b) [O. lambertii], *Rosa blanda* [R. woodsii], *Musenium* [*sic*] *tenuifolium, Applopappus armerioides* [Stenotus armerioides] and *nuttallii* [Xanthisma grindelioides], *Townsendia grandiflora, Hymenopappus* sp. (undoubtedly *H. luteus* Nutt., see No. 196 [*sic*, 195]) [Hymenopappus filifolius var. polycephalus], *Actinella acaulis* [Tetraneuris acaulis], *Cnicus pitcheri* [Cirsium undulatum], *Gilia spicata* [Ipomopsis spicata], *Chenopodium* sp. (No. 321) [C. fremontii], *Eurotia lanata* [Krascheninnikovia lanata], *Eriogonum annuum, Hedeoma drummondii, Agropyrum tenerum* vars. (two varieties 468 and 470 of my list) [Elymus trachycaulus], *Zygadenus* sp. in fruit [Zigadenus venenosus var. gramineus], *Juncus tenuis* var. (No. 379b) [J. tenuis].

In the meantime another cloud had risen, and down came a heavy shower. Although we had taken umbrellas, we became thoroughly soaked before we came home.

July 7. — Rains the larger part of the day. We looked over the presses, examined some plants, wrote labels, etc., but did not do any collecting. In the evening it cleared up. We hitched on the horses and drove to Mr. N[icholas] Frostrom's, partly in order to be nearer the Lawrence Fork, which we intended to visit.

July 8. – From Mr. Frostrom I heard that Rev. Bergquist had left some mail for us at Mr. Johnson's, one of the neighbors. We drove over there in the morning, and now I received my commission, which had been sent from Lincoln to me under the address of Rev. Bergquist. From there we drove down into the Lawrence Fork. This is a valley with a big sand-draw and several springs, which form a brook. This runs for the greater part in the sand, but here and there it comes up into the light.

On the hills surrounding the Lawrence Fork,

we collected the following: *Lesquerella ludoviciana* [Physaria ludoviciana], *Arenaria franklinii* var. (No. 32 of my list) [Eremogone hookeri], *Astragalus caespitosus* [A. spatulatus] and *triphyllus* [A. hyalinus], *Lathyrus ornatus* [L. decaphyllus], *Ribes cereum* [R. cereum var. inebrians], *Ribes aureum* [R. odoratum], *Paronychia jamesii*, *Eurotia lanata* [Krascheninnikovia lanata], *Eriogonum alatum*, *E. flavum*, *Andropogon scoparius* [Schizachyrium scoparium], etc. In the valley we found, *Corydalis curvisiliqua* [C. curvisiliqua subsp. occidentalis], *Polanisia trachysperma* [P. dodecandra subsp. trachysperma], *Cleome integrifolia* [C. serrulata], *Thermopsis rhombifolia*, *Psoralea campestris* [Pediomelum digitatum] and *lanceolata* [Psoralidium lanceolatum], *Petalostemon* sp. (No. 58) [Dalea candida var. oligophylla], and *tenuifolius* (No. 59) [Dalea purpurea var. purpurea], *Astragalus adsurgens* [A. laxmannii], *Prunus demissa* [P. virginiana] and *pumila* [P. pumila var. besseyi], *Rosa woodsii*, *Epilobium adenocaulon* [E. ciliatum], *Oenothera serrulata* and *albicaulis* var. (No. 112), *Symphoricarpos occidentalis*, *Crysopsis* [*sic*] *villosa* var. *canescens* [Heterotheca villosa], *Artemisia frigida*, *Asclepias speciosa*, *Physalis* sp. (No. 273) [P. hispida], *Penstemon glaber*, *Aphyllon ludovicianum* [Orobanche ludoviciana], *Monarda fistulosa* var. *mollis* [var. menthifolia] and *citriodora* [M. pectinata], *Abronia fragrans*, *Chenopodium Fremontii* var, *incanum* [C. fremontii], *Polygonum ramosissimum*, *Elymus canadensis*, *Agropyrum glaucum* var. (No. 464) [Elymus lanceolatus], *Poa nevadensis* (?) (No. 458) [P. secunda] *Oryzopsis micrantha*, *Panicum virgatum* var. *confertum* [P. virgatum], *Munroa squarrosa*, and *Solanum triflorum*. The last two seem to be confined to the prairie-dog towns.

The prairie dogs seem to disappear more and more, as the land becomes broken. In Lawrence Fork they were still numerous. Here are also found prairie wolves, badgers, wildcats, etc. Last year I saw a porcupine caught in the Fork.

The Lawrence Fork has for years been the headquarters of ranchmen, on account of the water it furnishes the stock. Even this year, there was a large herd there. We did not dare to camp until we came to the house of a settler, who had taken a claim of preemption and had his land fenced in. Here we pitched our tent for two days.

July 9. – In the morning we walked up the Fork a couple miles to a canyon in which I, in 1890, had collected a *Euphorbia* which I had named *E. cuphusperma* [E. davidii]. Before we reached the place, we collected in the sand-draw *Krynitzkia crassisepala* [Cryptantha minima] and *jamesii* [Cryptantha cinerea var. jamesii], *Gerardia tenuifolia* [Agalinis tenuifolia var. parviflora], and *Euphorbia hexagona*. In the canyon I found about a dozen specimens of *Euphorbia cuphusperma* (No. 361) [E. davidii]. We also saw several specimens of *Acerates auriculata* [Asclepias engelmanniana], but only in bud. When returning home, we collected, among others, *Cucurbita foetidissima*, *Erigeron pumilus* and *divergens* [E. bellidiastrum], *Asclepias verticillata* var. *pumila* [A. pumila], and *Ipomaea leptophulla* [*sic*].

After dinner we went into the plum thicket that grew near the place of encampment. In and around the same we found *Glycyrrhiza lepidota*, *Astragalus canadensis*, *Geum album* [G. canadense] and *striatum* [*sic*, for strictum][G. aleppicum], *Echinospermum floribundum* [Hackelia floribunda], *Mertensia lanceolata*, *Physalis longifolia*, Nutt. (No. 272) [P. hispida], *Parietaria* [P. pennsylvanica], and *Scirpus pungens* [Schoenoplectus pungens].

In the upper part of Lawrence Fork, there are no trees except plums and cherry trees, *Prunus americana* and *demissa* [P. virginiana]. Of shrubs there are the sand-cherry, *Prunus pumila* [P. pumila var. besseyi], two currants, *Ribes aureum* [R. odoratus] and *cereum* [R. cereum var. inebrians], poison ivy, *Rhus toxicodendron* [Toxicodendron radicans var. rydbergii], *Rhus canadensis* var. *trilobata* [R. aromatica], and *Symphoricarpos occidentalis*. In the lower part of the Fork, there are several groves of woods. In the one we visited, there grew, besides the trees mentioned, also three willows, *Salix cordata* var. *vestita* [S. eriocephala], *longifolia* [S. interior], and *amygdaloides* (?), white elm, hackberry, boxelder, cottonwood, wild grapes, and Virginia creeper.

We then mounted the hills north of the place. Here, as well as the hills all over between and around Lawrence Fork and the Pumpkin Valley, grew scattered trees of the Rocky Mountain yellow pine, *Pinus ponderosa* var. *scopulorum*. On these hills we also found the following: *Linum rigidum* [L. compactum], *Astragalus pubentissimus* (No. 74) [A. lotiflorus] and *kentrophyta, Gilia spicata* [Ipomopsis spicata], *Euphorbia montana* [E. robusta], and *Bouteloua hirsuta*. Going down into a little valley on the other side, we collected *Lupinus argenteus* var. *procumbens* [L. argenteus] and *Delphinium azureum* var. (No. 8) [D. virescens]. This valley led to a small canyon, which we followed up to the tableland. Here we collected *Rhus toxicodendron* [Toxicodendron radicans var. rydbergii], *Cercocarpus parvifolius* [C. montanus], *Acerates viridifolia* var. *lanceolata* [Asclepias viridiflora (narrow-leaved form)], *Krynitzkia glomerata* [Cryptantha thyrsiflora], *Hedeoma drummondii, Polygonum tenue* [P. sawatchense?] and *lapatifolium* [*sic*] var. *incanum* [P. lapathifolium], etc. We returned at sunset.

July 10. – In the morning we broke up, drove down the Lawrence Fork. Here it turned toward the north, widened to a valley several miles wide. In this valley there is and has been for several years a big horse ranch, taking in not only half of the valley but also the hills for several miles around. The nearer we came to the place where the Lawrence Fork opens into the Pumpkin Valley, the more water did we find in the creek, and also the more trees growing around the same. Just opposite the last cliffs, the extremity of the tableland between the Lawrence Fork and the Pumpkin Valley, we stopped for dinner. We first mounted the cliffs, on the other side of which we found a deep canyon. On the cliffs we collected *Actinella acaulis* [Tetraneuris acaulis], *Astragalus caespitosus* [A. spatulatus], *triphyllus* [A. hyalinus], and *kentrophyta*, and *Musenium* [*sic*] *tenuifolium*; and in the canyon *Clematis ligusticifolia, Brickelia* [*sic*] *grandiflora, Senecio douglasii* [S. riddellii], and *Mentzelia ornata* [M. decapetala], the latter in bud only. We then botanized along the brook, and collected, among others, *Epilobium adenocaulon* [E. ciliatum], *Berula angustifolia* [B. erecta var. incisa], *Lactuca ludoviciana, Mimulus jamesii* [M. glabratus var. jamesii], *Veronica americana, Mentha canadensis* [M. arvensis], *Lycopus sinuatus* [L. americanus], *Glyceria* sp. (No. 461 of my list) [Poa arida], *Lemna minor*, and *Zanichellia* [*sic*] *palustris*.

After dinner we started again. We drove northward till we came into the valley and then westward. We saw some clouds rise in the northwest. We, therefore, intended to stop as soon as we came to a place where we could get water. We stopped at the first house, but we had scarcely unhitched the horses when it began to hail. We led the horses into a shed, covered the buggy, and ran into the house. Now came down hailstones as large as hen's eggs. The Pumpkin Valley, so bragged of by land agents, seems to be a rather unlucky spot, as it has not only suffered much from drought during the last years but also is frequently visited by hailstorms. This storm did not do very great damage where we were, but northwest of the place, viz. north of Freeport, the crop was altogether ruined.

Not being very late in the day, we concluded, after the storm was over, to continue our journey, and, if possible reach the Hackberry Springs. Around these there is a valley, which in 1890 was very rich in plants. The Hackberry Springs have in former times been the headquarters of ranch men but had been vacated for some years. This year a rancher had taken his abode there again. When reaching the place, we became very disappointed. Everything, except the big trees, was eaten up or tramped down. We did not even dare to pitch our tent in the neighborhood of the big herds, therefore we drove a few miles up on the hills, where we camped.

July 11. — In the morning we drove down to the springs to get water. The woods consisted of hackberry, boxelder, plum, white elm, green and red ash, of which latter we found a tree that measured over 7 feet in circumference. We then drove a few miles in a sand-draw and botanized in the same. We collected among others *Lupinus argenteus* var. *procumbens* [L. argenteus] and *Eriogonum annum* [*sic*]. We then returned to the place where we had left our tent, picketed the horses, and ate our dinner.

In the afternoon, we went down into the canyons north of the place. Here we found *Ribes cereum* [R. c. var. inebrians], *Cucurbita foetidisssima* H.B.K., *Musenium* [*sic*] *tenuifolium, Galium boreale, Oxybaphus nyctaginius* [*sic*] var. *longifolius* [Mirabilis hirsuta], *Stipa comata*, etc.

In the evening, we drove across the divide, 3 or 4 miles, to Mr. N[icholas] Frostrom's, where we stayed over Sunday.

July 12. — Sunday, — no botanizing.

July 13. — We drove down into the Pumpkin Valley again, botanizing in the lower part of the same canyon we had visited on July 6. Here we collected, among others, *Ranunculus cymbalaria*, *Musenium* [*sic*] *tenuifolium*, *Brickelia* [*sic*] *grandiflora*, *Senecio canus* var. [Packera cana], *Gilia spicata* [Ipomopsis spicata], *Monarda fistulosa* [M. f. var. menthifolia], *Abronia fragrans*, and *Sisyrinchium anceps* [S. montanum].

In the afternoon we drove across the valley to Freeport, a city consisting of a store, a post office, a living house, and a few empty houses. Here we pitched our tent for two days.

July 14. — In the forenoon, we had all we could do in pressing and changing paper in the presses. In the afternoon we started for the hills north of the Pumpkin Valley. These were of the same appearance as those of the south side, only, perhaps, a little steeper and more barren. We found here the same plants as on the Courthouse Rock, and, besides those, *Penstemon glaber*, two low tufted plants and a low stunted brush, neither in bloom. The first two I afterwards recognized as *Gilia pungens* var. *caespitosa* [Leptodactylon caespitosum] and *Phlox hoodii*, the latter as *Bigelovia howardii* [Ericameria parryi var. howardii]. We walked around for a long while, without finding anything new. We intended to return, but wished first to take a view over the country, wherefore we mounted what we thought was the highest hill. Ascending the same, we suddenly struck a patch of *Astragalus pictus* var. *filifolius* [A. ceramicus var. filifolius]. Only three specimens in bloom were collected, but we gathered a good supply of the beautiful mottled pods. When we reached the top, we saw that we were not on the highest hill, but that this lay about 40 rods further northwest. We went up there and, indeed, we saw a view worth seeing: south of us was the Pumpkin Valley, 6 or 7 miles wide, extending east and west, bordered by hills on the south; in the valley, the winding creek, small farms here and there, and the Courthouse Rock in the eastern opening. On both sides of us, east and west, there was the ridge, with the hills covered with scattered trees of the Rocky Mountain yellow pine, and rising into the Wildcat Hills in the west: North of us was the North Platte Valley, the river with its small islands in the middle surrounded by green meadows, between the river and us, several isolated buttes, the Scotts Bluff in the northwest and the Chimney Rock in the east, and as a background, the hills north of the Platte. The only things lacking to make the picture perfect, were thrifty towns and well-built farm houses surrounded by smiling groves and orchards. I hope they will come by and by.

In going down, we followed a canyon so narrow that, in many places, we could touch both sides thereof at the same time by stretching out our hands, and still over 100 feet deep. We found here unusually large specimens of *Lygodesmia juncea* and *Juncus bufonius*, and, on a hill below, *Arabis holboellii* [Boechera grahamii] in fruit. Before we reached Freeport it began to rain. It also rained during the night but only small showers.

July 15. — In the forenoon we broke up. We intended to move our camp to Ashford. From Freeport to Ashford is 16 miles. We followed the post road, running not far from the creek. The soil of Pumpkin Valley is sandy but, in many places, far from poor, but not as rich as the tableland south of the valley. Along the creek were meadows with quite good grass, but here and there the soil was somewhat alkaline. *Distichlis spicata* var. *stricta* [D. spicata] and *Suaeda diffusa* [S. calceoliformis] grew in several places. On the way we collected *Heliotropium curassivicum* [*sic*] [H. curassavicum var. obovatum] and unusually large specimens of *Psoralea argophylla* [Pediomelum argophyllum], *Gaura coccinea*, and *Oenothera albicaulis* [Oe. nuttallii]. At noon, we drove up to the hills, which were much steeper, and much whiter than I have seen anywhere else in the west. They were white as chalk and had no vegetation except in the crevices. The top of these as well as of all the hills in the region was nearly flat. A hailstorm had shortly before passed thereover, and we could see marks from the hailstones in the rock. The only plant we found worth mentioning was *Gilia iberidifolia* [Ipomopsis congesta]. Below we collected *Galium boreale*, *Ribes cernuum* [R. c. var. inebrians], and *aureum* [R. odoratum], etc. Between these hills and the Wildcat Hills there was a valley, 5 or 6 miles wide and extending about as many miles northward into the divide. At the foot of the Wildcat

Hills is situated Ashford, another little town of a dozen houses, of which two were stores. Ashford had had a newspaper of its own, but the printers moved away the day after our arrival to town. I, maybe, was the last one that visited the office in newspaper business when I bought for 15 ¢ papers to put plants in.

July 16. – We mounted the cliffs above Ashford. These were very steep and hard to climb, about 300 to 400 feet high. In the crevices grew a few stunted pines. We also found the usual cliff plants, all those that grew on the Courthouse Rock and on the hills north of Freeport, and, besides these, *Happlopappus nuttallii* [Xanthisma grindelioides] and *Hymenopappus* sp. (*H. luteus* Nutt.?, No. 195 of my list) [Hymenopappus filifolius var. polycephalus]. When we reached the top, we saw that on the other side of a small valley, there were several buttes, lying as if thrown on the top of the ridge of hills. They were between 100 and 200 feet higher than any hill in the neighborhood. The Scotts Bluff is claimed to be the highest point in Nebraska. We saw the Scotts Bluff from here, but it seemed to us as being lower. From the Scotts Bluff, we also saw the Wildcat Hills, and they seemed to be higher.

Just as we reached the top of the nearest of these buttes, I found a glandular pruinose variety of *Sisymbrium canescens* (No. 17 of my list) [Descurainia pinnata]. When going down we were struck by the strange form of the red cedars. The trees were naked below and had a rounded top. The leaves were shorter and the berries smaller than usual (See also my list, No. 474, where I have given it nearly half a page) [Juniperus scopulorum]. Here the bluffs formed a steep incline down into the valley. The sides were covered by cedars and pines. In the canyon below grew boxelder, ash, and hackberry. In the sand-draw we collected *Rosa woodsii*, *Mentzelia nuda*, *Krynitzkia fendleri* [Cryptantha fendleri], etc. The sand-draw led down to the Pumpkin Creek. Here, in some places, grew quite good grass, but in others the soil was alkaline and the ground was covered with *Suaeda diffusa* [S. calceoliformis] and *Distichlis spicata* var. *stricta* [D. spicata]. In the creek we collected *Sagittaria variabilis* [S. cuneata], *Epilobium adenocaulon* [E. ciliatum], *Zannichellia* [Z. palustris], and a *C[h]ara*.

July 17. — We drove 5 or 6 miles up the creek. The land seemed to become more and more alkaline. We stopped and botanized in a meadow. Here we collected, among others, *Crepis runcinata*, *Antennaria plantaginifolia* [A. microphylla] *Arabis* sp. (No. 13 of my list) [Thelypodium integrifolium], *Suaeda diffusa* [S. calceoliformis], *Triglochin maritima*, and a *Plantago* near *P. eriopoda* (No. 304 of my list) [P. eriopoda]. When driving home we were caught in a thunderstorm, the horses became scared and hard to manage, and we were quite wet before we reached Ashford.

July 18. – In the morning we packed all our dry plants in my trunk and left it in charge of the storekeeper at the place until we should return. As the stage road through Ashford was the only road that led over the hills to Gering, from this part of the valley, we expected to take the same road back. In the morning we broke up. The stage road led through the valley mentioned above, for about 4 or 5 miles, whereafter it ascended the hills. Here it wound in nearly every possible direction until the top of the ridge was reached, the road being at least three times as long as the straight line from the valley to the top. The descent was more straight, but also more steep. We had to make good use of the brakes on the wheels. The road followed one branch of the Cedar Canyon. The other branch was spanned by a bridge. A little below this bridge, we unhitched our horses and tied them out to graze while we took our dinner. After dinner we botanized in the canyon and collected *Lupinus argenteus* var. *procumbens* [L. argenteus], *Cercocarpus parvifolius* [C. montanus], *Lepachys columnaris* var. *pulcherrima* [Ratibida columnifera f. pulcherrima], *Andropogon hallii* [A. gerardii susp. hallii], and *Chrysopogon nutans* [Sorghastrum nutans].

In the afternoon we reached Gering, the County Seat of Scotts Bluff County. We drove to the post office to hear if we had any mail. We met Mr. Richardson, a prominent lawyer, and Mr. Wood, the editor of the *Courier*, with whom we had a long talk.

When the good people of Gering saw our wagon with presses, collecting cases, etc., they imagined that we were in some way connected with the survey of the irrigation ditch. Irrigation was the burning question of the day in Scotts Bluff Co. I have been surprised more than once during the trip over all the strange questions we have received concerning our work. The people, I saw, were just as much surprised when seeing men sent out to that arid region for the purpose of picking flowers and gathering weeds. In Sweden, when a botanist came with a collecting case over his shoulder, nearly everyone knew what his business was.

After we had bought some provisions, we pitched our tent on the prairie, about half a mile west of town.

July 19. — Sunday, no botanizing.

July 20. – Gering is situated about two miles east of Scotts Bluff. Very naturally, the Bluff would be the first place to visit. Early in the morning, before sunrise, we were up. When we had eaten our breakfast and watered our horses, the fog had not yet risen, so we did not see the Bluff, but we walked in the direction of it. We suddenly struck a canyon running west-southwest and east-northeast. On the sides grew *Atriplex nuttallii* [A. canescens]. We went down into the canyon, where we found *Pseudocymopterus anisatus* Coulter & Rose [Lomatium nuttallii]. When we came near the head of the canyon, the fog had dispersed, and we found ourselves not very far from the mountain, being on the south side of it.

Scotts Bluff is said to be 750 feet above the level of North Platte River at the place, and the highest point in Nebraska, or nearly 6000 feet above the sea. For my part, I believe that the Wildcat Hills, mentioned above, are still higher. If the Scotts Bluff is 750 feet above the North Platte, it is over 600 feet higher than the ground at its base. It is said that the bluff cannot be mounted except at two places, one on the south side, and one on the west side. We saw a place where the ground seemed to rise gradually at least to near the top, and we began the ascent. When about 60 or 70 feet from the top, we found that it was impossible to go any further, there being a perpendicular wall 30 or 40 feet high. We sat down for awhile in the shade of the cliff to take a rest. My eyes happened to strike a peculiar *Physalis*, which I have named *P. mollis* var. *cinerascens* [P. hederifolia var. comata]. Among the rocks we also collected *Viola nuttallii*, *Dalea aurea*, and *Chenopodium leptophyllum* var. *subglabrum* [C. pratericola]. We then went around the southeast corner of the mountain. Here the cliffs were perpendicular for a height of about 400 feet. On the side I found a *Phacelia* (No. 247 of my list) [P. hastata]. Right opposite Gering, the bluff rose gradually at an angle of about 60°; here we found the east path to the top. Near the top, it had been necessary to cut steps in the rock. The view from the top of the Scotts Bluff was magnificent. On the south and west, we had the Cedar and the Mitchell valleys, separated by a row of bluffs a little lower than the one we stood upon, and bordered to the south by the ridge we had crossed two days before. North of the bluff, we saw the North Platte River, which winds its way through Scotts Bluff Co. from the northwest to the southeast corner. East of us we had the town of Gering, and in the far west, the Laramie Peak [in Wyoming], the top of which is 9000 feet over the sea. We returned very tired to our camp. In the evening, we tended to the presses.

July 21. — We went up to the canyon, a part of which we had followed the preceding day, but followed it downwards this time. We found two varieties of *Ribes aureum* [R. odoratum], one with black and one with yellow berries. The black berries have a taste much like that of the berries of *Ribes nigrum* and *floridum* [R. americanum]; the yellow berries are more pleasant and slightly sourer. I did not see any intermediate forms, although both were common in Scotts Bluff County. I think that they have a good right to be separated as well-marked varieties. We also collected *Malvastrum coccineum* [Sphaeralcea coccinea], *Petalostemon* sp. (No. 58 of my list) [Dalea candida var. oligophylla], *Astragalus adsurgens* [A. laxmannii], *Bigelovia graveolens* var. *glabrata* [Ericameria nauseosa var. graveolens], *Phlox douglasii* var. (No. 238) [P. andicola], *Krynitzkia glomerata* [Cryptantha thyrsiflora], *Amarantus* [*sic*] sp. (No. 313) [A. blitoides], *Coriospermum hyssopifolium* [Corispermum americanum], etc.

The canyon ran down to the North Platte River, which we followed for a couple miles. On the bottomland, we collected, among others, *Potentilla anserina* and *supina* [P. paradoxa], *Geum strictum* [G. aleppicum], *Aster foliosus* [?], *Iva axillaris*, *Eustoma russellianum* [E. grandiflorum], *Calamagrostis longifolia* [Calamovilfa longifolia],

Spartina cynocuroides [*sic*][S. pectinata], *Elymus canadensis*, *Fimbristylis spadicea* var. *castanea* [F. puberula var. interior], *Agropyrum glaucum* var. (No. 463) [Elymus smithii], and *tenerum* var. (No. 471) [Elymus trachycaulus]. On the islands of the North Platte, we found *Shepherdia argentea* and a variety of *Ribes aureum* [R. odoratum] with big, ellipsoid, black fruits. When we returned across the prairies, we found *Astragalus pectinatus* in fruit.

July 22. – South of Scotts Bluff proper there is a ridge of bluffs a little lower. Between the east-northeast end of this ridge and the Scotts Bluff, there is a narrow pass through which the old California trail runs, the land between the Bluff and the [North] Platte River being too much broken by canyons to make a road possible. We intended to botanize on the western side of the Bluff. When we approached the pass, we found several big clumps of *Pseudocymopterus anisatus,* Coulter & Rose [Lomatium nuttallii]. When on the other side, we picketed our horses and ascended the foot-hills. Here Mr. Flodman found *Stephanomeria minor* [S. runcinata], taking it for a *Lygodesmia.* I had taken, on the east side of the mountain, one small specimen of the same, all covered with rust. I took it then for *Lygodesmia juncea,* believing that the pale pink color was due to the work of the rust. It was not before I gave it a closer examination at Lincoln that we found that we had been mistaken when referring it to the genus *Lygodesmia.* We also collected *Astragalus kentrophyta* and *triphyllus* [A. hyalinus], *Erigeron canus* and *divergens* [E. bellidiastrum], *Gilia pungens* var. *caespitosa* [Leptodactylon caespitosum], and *iberidifolia* [Ipomopsis congesta]. When returning home we collected, in the pass, *Rosa woodsii* and *Cnicus pitcheri* [Cirsium undulatum]. Before, we had collected, in a canyon running around the northwest corner of the Bluff, *Oenothera caespitosa, Mentzelia ornata* [M. decapetala], *Brickelia* [*sic*] *grandiflora, Lepachys columnaris* var. *pulcherrima* [Ratibida columnifera f. pulcherrima], and *Euphorbia petaloidea* [E. missurica var. petaloidea]. In a canyon running down from the bluffs north of the pass, I found a crucifer with the leaves crowded at the base in the same manner as in the cabbage or the house leek. It was unusually large specimens of *Physaria didyma* [*sic*] [P. brassicoides].

July 23. — In the morning, we walked to the hills south of the Scotts Bluff. Here we botanized on the sides of the hills and in the canyons below, and found among others, *Erysimum asperum, Lesquerella ludoviciana* [Physaria ludoviciana], *Physaria didyma* [P. brassicoides], *Arenaria franklinii* var. (No. 31) [Eremogone hookeri], *Oxytropis lambertii* var. *sericea* [O. sericea], *Aster tanacetifolius* [Machaeranthera tanacetifolia], *Apocynum canabinum* [A. cannabinum], *Phlox caespitosa* var. *rigida* [P. hoodii], *Aphyllon fasciculatum* var. *luteum* [Orobanche fasciculata], *Eurotia lanata* [Krascheninnikovia lanata], *Yucca angustifolia* [Y. glauca], and *Fritellaria* [*sic*] *atro-purupurea* [Fritillaria atropurpurea]. Mr. Flodman climbed the bluff and walked around for a while on the top. When he descended again, I sat down to wait on him. He, however, passed without noticing me. Neither did I notice him and began to halloo, but received no answer. I continued eastward and went around the extremity of the ridge, an isolated small butte. Here I collected *Bigelovia douglasii* [Chrysothamnus viscidiflorus] and *Oenothera caespitosa.* Going home, I followed a canyon through the prairie. Here I found *Asclepias auriculata* [A. engelmanniana]. When I arrived home, Mr. Flodman was there and had been there for about three hours. I had my dinner after 3 o'clock that day. In the evening, we tended to the presses.

July 24. — The field of our botanizing for the day was the badlands north of the Scotts Bluff. Surely, "badlands" was a very appropriate name, because from the Bluff clear down to the [North] Platte River, there was a piece of land, several sections, all made up of canyon after canyon winding down and separated from one another by narrow steep ridges. Not a green spot was seen. We first botanized near the bluff. We wished to mount one of the foothills. Mr. Flodman tried first but had to give up. I, however, succeeded by crawling on my knees and elbows, placing as much as possible of my arms and legs to the smooth rock, in order to make the friction greater. I regarded myself well paid for my hard labor, when I, near the steep wall of the mountain, in the shadow of some pines, found a *Mentzelia,* near to *albicaulis* and *dispersa* (No. 121) [M. albicaulis]. When I had come down, we followed one of the canyons downward. The canyon was winding around, so I am sure we walked over 6 miles before we reached the river, although the distance from the mountain to the North Platte is not over two miles. In the canyon and on the sides, we collected *Oenothera biennis* var. (No. 108), *Mentzelia ornata* [Mentzelia decapetala], *Gutierrezia euthamiae* [G. sarothrae], *Physalis* sp. (No. 273 of my list) [P. hispida], *Muhlenbergia pungens*, and *Sporobolus cryptandrus* var. (No. 433) [S. cryptandrus].

At the mouth of the small stream that ran in the canyon, there was a small strip of bottomland along the river. Here we found *Petalostemon* sp. (No. 58b) [Dalea candida var. oligophylla], *Vicia americana* var. *linearis* [V. americana var. minor], *Solidago serotina* [S. gigantea], *Helenium autumnale*, etc. We took a bath in the North Platte, whereafter we started homeward, following the river.

The ridges between the canyons stooped suddenly into the river, leaving no bottomland at all. We had to cross ridge after ridge and canyon after canyon and were very tired when we reached our camp.

July 25. – We drove to a canyon, the Carter Canyon, 6 or 7 miles S.W. of Gering. When we approached the canyon, we saw some dark green trees with a conical crown. I took them at first for *Juniperus virginiana*, which was the only tree I knew of that resembled these in darkness and shape. I found, however, that they were not cedars at all but poplars [P. ×acuminata] near to *Populus balsamifera* and *angustifolia*. This poplar differs from the former in having the leaves green on both sides and from the latter in having much longer petioles and larger teeth on the leaves. A fuller description is given in my list, on page 30. Besides this poplar there grew in the canyon the following trees and shrubs: *Populus angulata* [P. deltoides var. monilifera], two varieties of *Salix longifolia* [S. interior], *Negundo aceroides* [Acer negundo], *Fraxinus viridis* [F. pennsylvanica], *Ulmus americana*, *Celtis occidentalis*, *Symphoricarpos*, and *Rhus toxicodendron* [Toxicodendron radicans var. rydbergii], and, on the surrounding hills, *Pinus ponderosa* var. *scopulorum*, *Juniperus virginiana* [J. scopulorum], and *Rhus canadensis* var. *trilobata* [R. aromatica]. Of herbs, we found a *Chenopodium*, which I regarded as new, (No. 321 of my list) [C. fremontii], *Ch. album* [C. berlandieri], *Epilobium adenocaulon* [E. ciliatum], *Kuhnia eupatorioides* var. *corymbulosa* [Brickellia eupatorioides var. corymbulosa], *Solidago missouriensis* var. (No. 153) [S. missouriensis], *Aster commutatus* [A. falcatus subsp. commutatus?], *Campanula rotundifolia*, *Sporobolus asperifolius* [Muhlenbergia asperifolia], *Alopecurus aristulatus* [A. aequalis], two varieties of *Agropyrum tenerum* [Elymus trachycaulus], and *Juncus bufonius*.

The brook in the canyon had been swelled up, and the water had been high over the banks but had sunken again. A hailstorm had passed over the region two days before. We found even, among the fallen leaves, hail-stones that had not melted yet.

In the afternoon we returned to Gering.

July 26. — Sunday, no botanizing.

July 27. — In the morning, we packed our wagon, bought some provisions, and broke up. We intended to enter the Kiowa Valley in the southwestern part of Scotts Bluff Co. The road to this valley leads through the Cedar Valley and then through an opening between the hills named the Roobadoo [Robidoux] pass (I do not know the spelling of the name but have spelled it as it is pronounced.). I had met with a Mr. Wells, who lives in the Kiowa Valley. He had described this road and invited us to come to his house. When we came to the other side of the Roobadoo, we stopped for dinner. After dinner, I went to a little hill at the side of the road. Here I collected *Eriogonum brevicaule* and *Astragalus triphyllus* [A. hyalinus], of which the former is new to Nebraska. It now began to rain. We crawled under the wagon after having covered it in order to protect the presses. When the worst shower was over, we started again and drove into the valley.

The land was drier than in both the Cedar and the Pumpkin valleys, and somewhat sandy, and here and there a little saline. We collected, on the way, *Atriplex argentea*, *Solidago nemoralis* var. *incana* [S. mollis], *Soldago missouriensis*, and a variety of the same (No. 154 of my list), *Achillea millefolium*, *Lippia cuneifolia*, etc. When we came near to Mr. Wells', we found *Eriogonum microthecum* [E. effusum var. effusum], also new to Nebraska.

July 28. — In the morning, Mr. Wells was preparing himself to go to Alliance with a load of bones, and, in the return, take a load of groceries for the store-keeper in Gering. From Mr. Wells' to Gering was 18 miles, and from there to Alliance, 50 miles. It had taken Mr. Wells and his son two days to pick the bones, and he had to haul the load 68 miles to market, and, then, he expected to get, at the very highest, $10. Mr. Wells had not had a crop for four years and had been obliged to sell what little he expected this year when still on the field. To pick bones and do a little hauling were the only means by which he, as well as many other settlers, could earn a little cash for their sustenance. Those that had been prudent enough to invest their money (if they had any to invest) in cattle had been able to get along better.

After breakfast, we drove up to the hills and entered a small canyon. Here I collected *Draba Holboelii* [Boechera grahamii], *Oenothera biennis* var. (No. 108), *Artemisia cana, Eriogonum brevicaule,* and *Euphorbia petaloidea* [E. missurica var. petaloidea], while Mr. Flodman drove down to find a place where he could cross the canyon with the wagon. He had to drive a couple miles before he could do so. As Mr. Wells had no well, but hauled the water for his animals, we had given our horses only a little water in the morning, knowing that there were several springs in the canyons. The horses needed water. There were no springs in the canyon I visited. We crossed two other small canyons without finding water. We crossed a third canyon, a little larger. When driving down on the other side of the same, we found a pool from which ran a small stream. Here we watered our horses, but had to carry the water up a steep bank about 50 feet high. Driving down a little further, we found that the canyon joined a still larger one, which we followed upwards, driving on the bottom. A mile or so up the canyon there was plenty of water, a stream, and several pools. Here we picketed our horses, took our dinner, and botanized in the afternoon. We collected, among others, *Epilobium adenocaulon* [E. ciliatum], *Menzelia nuda* [M. nuda var. stricta], *Berula angustifolia* [B. erecta var. incisa], *Erigeron canadense* [Conyza canadensis], *Cicuta* sp. (No. 264) [C. maculata], *Cycloloma platyphyllum* [C. atriplicifolium], *Eriogonum annum* [*sic*], *Euphoria serphyllifolia, Scirpus lacustris* [Schoenoplectus tabernaemontini] and *atrovirens* var. *pallidus* [S. pallidus], *Panicum crus-galli* [Echinochloa muricata var. microstachya?], and *Eatonia* sp. (No. 453 of my list) [Sphenopholis obtusata var. obtusata]. The woody flora consisted of elm, hackberry, cottonwood, green ash, willows, red cedar, Rocky Mountain yellow pine, wild grapes, Virginia creeper, dogwood (*Cornus stolonifera* [C. sericea]), *Ribes aureum* [R. odoratum], *cereum* [R. c. var. inebrians], and *floridum* [R. americanum], *Rhus toxicodendron* [Toxicodendron radicans var. rydbergii] and *canadense* var. *trilobata* [R. aromatica].

In the evening we drove out on the prairie in order to reach some homestead. Our supply of bread was nearly out. We reached a house, but we could not buy any bread. We put our tent up and stopped overnight.

July 29. – In the morning we spent a few hours in repairing collars, harnesses, our clothing, etc., whereafter we drove southward into the Pleasant Valley, which is an extension of the Pumpkin Valley and separated from the Kiowa Valley by a sandy ridge, which ends in a small butte, the Eagle Nest. We saw a house with a square front as stores used to have. We drove there in order to buy, if possible, bread or crackers. We were out of bread, and had crackers enough for a day or so. The store was vacated. We therefore continued to drive until we reached a farmhouse. We could not get any bread, but the farmer's wife promised to bake us some biscuits in the morning. She did so, but the biscuits were not very palatable, because the lady used, at least in my opinion, too large a quantity of lard in them.

We went up to a ridge of hills, the "76" Mountains [*sic,* actually Sixtysix Mountain, adjacent in Goshen Co., Wyoming (Urbanek, 1988, and Wyomingplaces.org/)]. A ranchman [Philip Dater], who had his headquarters among these hills, had his cattle marked with the number 76 [66]. Hence the name. The [state] line is running along this ridge. We went even a few rods into that state. On these hills we collected *Oenothera hartwegii* var. *lavandulaefolia* [Oe. lavandulifolia], *Musenium* [*sic*] *tenuifolium, Galium boreale, Artemisia canadensis* [A. campestris subsp. caudata], *Phlox hoodii, Gilia pungens* var. *caespitosa* [Leptodactylon caespitosum], etc. When we went home, we came down into a small canyon in which were a few springs.

These gave rise to a small brook, which sank down into the sand when the valley was reached. Along the stream we collected *Ranunculus cymbalaria*, *Psoralea argophylla* [Pediomelum argophyllum], and a variety of the same with green leaves (No. 50) [a form of Pediomelum argophyllum], *Euphorbia glyptosperma* and *petaloidea* [E. missurica var. petaloidea], and, upon the prairie below, *Astragalus pectinatus* in fruit.

July 30. – Raining the whole day. If judging from the showers we met with during our stay in western Nebraska, we should have come to the conclusion that western Nebraska was a land of rain rather than of drought. In the morning, we turned over the presses, labeled some plants, etc. The farmer and his son came over to our tent. The farmer was an old soldier and related stories of the Civil War. At noon, he invited us to his house to take dinner with him, which we did. He spoke about the crop. He had raised very little since he settled on his homestead on account of the drought and hailstorms. One of the latter had partly destroyed his wheat this year, but his corn looked promising. We had seen his cornfield on the preceding day, and it was the first we had seen in the county hitherto. His fields were the best cultivated we had seen in western Nebraska. He claimed that it was hard to keep the fields clean on account of the "mountain purslane," as he named it. In the afternoon, between two showers, I followed him out to the field to see what the mountain purslane was, and found that it was an *Amarantus* [*sic*], very much resembling *A. blitoides*, but prostrate and having 5 stamens. It is No. 313 of my list [Amaranthus blitoides]. I also collected some specimens of *A. blitoides* for comparison sake.

July 31. – In the morning we broke up, stopped an hour or two at the Eagle Nest, collecting, but did not find anything of interest. We then drove northward and passed a house where we tried to buy bread, but in vain. We were now half a mile from the Wyoming line. When watering our horses at the well, we found growing near it a *Nasturtium*, which I have named *N. curvisiliqua* (No. 20) [Rorippa sinuata]. We followed a road running northward, but this road ended at an abandoned stone quarry. We then had to drive across the rolling prairie. We were obliged to drive mostly in walk [*sic*], partly because the ground was so uneven, partly because it was hard for the horses to avoid stepping in big clumps of cactus that grew on the hills. The soil was very sandy, the grass both thin and short. As we crossed we found *Bahia oppositifolia* [Picradendriopsis oppositifolia] and *Chenopodium fremontii* var. *incanum* [C. fremontii]. Toward evening we came to a settled region again. We succeeded in getting bread and were shown the road to the mouth of Horse Creek, where we intended to stop. Before we reached the Horse Creek, we had to cross the new irrigation ditch. There was no bridge made yet. The horses were afraid of going down the steep bank. Mr. Flodman jumped off and led the horses by the bridles. A little before sunset we reached the horse ranch at the mouth of Horse Creek, and here we camped.

August 1. – The ranch lay between the Horse Creek and the [North] Platte River. The ranch proper had been moved away, only about 150 horses were left in charge of one man. The houses were empty except one. I read a sign over the door of one of the houses. It read as follows:

NOTICE TO TRAVELERS
At all ranches of this company,
travelers stopping at the ranches,
are charged at the following rates:
One meal..................50¢
One horse, pr feed.....25¢

When a person knows what kind of board was kept at the ranches, and that the "feed" meant simply a bundle of hay, he knows that these prices were unreasonably high. They were made in order to keep the "cow-punchers" from the ranches in the wintertime. The "cow-punchers," however, were generally on good terms with the cooks, and the man in charge of the ranch now said that, during the five years he had been on the ranch, the owner did not receive more than 75¢.

The water supply for the household was taken from a spring near the houses. This spring gave rise to a stream, which emptied into the [North] Platte River about 3/4 of a mile below the ranch. Along this stream was the first place we chose for our botanizing. We saw a heron sitting on a fence post, about 40 rods from the stream. Mr. Flodman went for his gun, an old musket, shot at the bird, but did not hit it. He also shot at a duck, but in vain. In the meantime, I botanized along the stream and collected *Ranunuculus cymbalaria* and *R. sceleratus*, *Potentilla anserina* and *supina* [P. paradoxa], *Epilobium adenocaulon* [E. ciliatum], *Grindelia squarrosa*, *Aster fremontii* [?] and *foliosus* [?], *Cnicus altissimus* var. *philipendula* [Cirsium flodmanii], *Teucrium occidentale* [T. canadense var. occidentale], *Mentha canadensis* [M. arvensis], *Scutellaria galericulata*, *Physostegia parviflora*, *Juncus nodosus*, *Lemna perpusilla* (No. 386 of my list) [L. minor], and *Sagittaria variabilis* [S. cuneata]. When reaching the mouth of the stream, I followed the Platte upwards and found here *Nasturtium palustre* [Rorippa palustris], *Strophostyles pauciflorus* [S. leiosperma], *Rumex salicifolius* [R. triangulivalvis], *maritimus* [R. fueginus], and another species intermediate between the two (No. 341 of my list) [R. crispus?], *Polygonum muhlenbergii* [P. coccineum], *Salix longifolia* and variety of the same (No. 369) [S. interior], *Fimbristylis spadicea* var. *castanea* [F. puberula var. interior]. Mr. Flodman, who had given up the hunt, had waded out to some of the islands and gathered, among others, *Helianthus maximiliani* var. (No. 138), *rigidus* [H. pauciflorus subsp. subrhomboideus], and *annuus*. He then joined me and we went along on the sands for a short distance. Here we collected *Gutierrezia euthamiae* [G. sartothrae], *Dys[s]odia chrysanthemoides* [D. papposa], *Erythraea douglasii* [Centaurium exaltatum], *Eustoma russellianum* [E. grandiflorum], *Gilia* sp. (no. 246) [G. pinnatifida], *Euphorbia glyptosperma*, *Cyperus aristulatus* [C. squarrosus], and *Sporobolus serotinus* var. (No. 436) [Muhlenbergia minutissima].

Going to the ranch, we crossed a part of the extensive meadows that extended as far as the eye could see upwards to the North Platte. A few miles west of us, there were big stacks of hay. The grass in the meadow was, at an average, two feet high, I shall enumerate a few of the grasses I noticed: *Panicum virgatum*, *Sporobolus airoides*, *Agrostis alba* [A. stolonifera], *Agropyrum glaucum* [Elymus smithii?], *Bouteloua oligostachya* [B. gracilis], *Elymus canadensis* and *Sitanion* [Elymus elymoides subsp. brevifolius], and, nearer the river, *Calamagrostis longifolia* [Calamovilfa longifolia], *Spartina cynocuroides* [*sic*] [S. pectinata], and *Phragmites communis* [P. australis]. Of these, *Panicum virgatum*, *Agropyrum glaucum* [Elymus smithii?], and *Bouteloua oligostachya* [B. gracilis] were regarded as the best hay grasses. I asked the man how he regarded *Agrostis alba* [A. stolonifera], which I showed him. He answered that the ranch-men regarded it as nearly good for nothing. He did not even know that it was a variety of the red top.

In the afternoon, we broke up. As it was Saturday, we tried to reach Gering before night. At last it became so dark that we could not see the trail. The old California trail was fenced up at several places, and the new trail was not driven enough yet. We came to a house and asked for the road. From the description that was given, we found that it was hard to follow, even in the day-light. We therefore put up our tent and camped over till Monday.

August 2. — Being Sunday, we rested from our work.

August 3. — We were about 5 miles from the Scotts Bluff in the Mitchell Valley. Not very much of the land was broken. Hitherto it had mostly been used for pastures, and, nearer the river for haylands. This year, the irrigation ditch had been completed. The cornfields had been irrigated once. The effect of that single watering, was plainly seen. The corn was the best we had seen this year. In a few years, the Mitchell Valley will be one of the very finest farming countries.

We followed the road for awhile, then drove down on the bottomland of the [North] Platte River. The grass was fine, about knee-deep. Along the river, from the badlands north of Scotts Bluff, as far west as the eye could see, we saw the big stacks of hay from the preceding year's crop. We botanized for a couple hours on the bottom. The following were in the collection we made: *Alisma plantago* var. *americana* [A. triviale], *Elymus sitanion* [E. elymoides subsp. brevifolius], *Liatris spicata* [L. lancifolia], *Solidago canadensis*, *Lactuca ludoviciana*, *Lobelia spicata*, *Erythraea douglasii* [Centaurium exaltatum], etc.

We stopped for dinner just west of Scotts Bluff and collected on the hills south of the road. We did not find anything of interest except some specimens of *Eriogonum multiceps* [E. pauciflorum var. gnaphalodes] with elongated pedicels, the involucre forming an umbell [*sic*] instead of a head. We arrived at Gering in the afternoon.

August 4. — We crossed the river and followed it upwards for 5 or 6 miles. Along the river there were good haylands. We also saw some good fields. The land was, and had been for two years under irrigation. We stopped near the river and took our dinner and botanized for a couple hours on the banks and on the islands, but did not find anything of interest. We then drove a few miles northwestwards, then turned north for half a mile, until we met a road running east. We followed this road and very soon crossed the irrigation ditch. On the other side, we came into a very sandy high prairie land. Here we drove miles after miles without seeing anything but low sunflowers, *Helianthus annuus* and *petiolaris*, and *Chenopodium leptophyllum* var. [C. pratericola]. To us it seemed as if we were driving through an old field, but evidently the ground had never been broken. Indeed, a miserable land. The only thing we collected was *Petalostemon macrostachyus* var. (no. 61) [Dalea cylindriceps]. We came out to a road running north just as a man from Snake Creek came hauling a load of wood from the divide between Pumpkin Valley and the Platte Valley, 75 miles from his home. We intended to drive up to the hills we saw north of us, but changed our mind when we heard that we could not get water within 15 miles from the place in that direction. We turned south until we came below the ditches, where we tented overnight.

August 5. – We took the main road running east and followed it for 6 or 7 miles, then turned south, until we reached the North Platte. Here we took a bath and ate dinner. Along the river were extensive pastures and haylands. A little higher up, we saw fine wheat and corn fields, the best I have seen in Nebraska. The heads of the wheat were very large and full of the plumpest grains. One farmer said that he had raised last year 80 bushels of oats to the acre. It pays to farm under irrigation. I think it would be a paying investment and a good deed, if the government could invest some hundred thousands of dollars in irrigation ditches in western Nebraska and other parts of the "Great American Desert," as the territory west of the Missouri River has been named.

When we crossed one of the smaller irrigation ditches, we broke our doubletree. We spliced it by tying a pole to it. We had to drive slowly to Gering. We found nothing new in the way of plants during the day.

August 6. — I walked up to Scotts Bluff to collect a specimen in bloom of *Bigelovia graveolens* var. *glabrata* [Ericameria nauseosa var. graveolens], and of *Dalea aurea*, which two we had found only in bud. In the meantime, Mr. Flodman had got a new doubletree made. Before we had everything packed and had bought the necessary provisions, it was near noon, wherefore we took dinner before we started. We arrived at Ashford at night.

August 7. — It was difficult to make room in the buggy for my trunk, which had been left at Ashford on July 18. This was done only by putting our clothing, the tent, etc., in the seat. The buggy was well loaded: my trunk and two cracker boxes, all full of dry plants, and a valise containing our provisions in the bottom behind the seat, the boxes and the valise set on end; on top of these, two satchels, one collecting case and five presses; under the seat a pack of drying paper, our cooking utensils, and the tent sticks; under our feet, the picket ropes and sticks, a hatchet, a water jug, etc.; in the seat our clothing and the tent; between our knees one collecting case; and under the wagon, the tent poles and the bucket.

We drove across the valley, following the stage road to Harrisburg. On the way, we found a *Physalis*, which is maybe *Ph. fendleri* (No. 270) [P. hederifolia var. comata]. A few miles from Harrisburg, we turned to the east, following a road running parallel to the hills on the south side of the valley. About noon we came to a settlement of Germans, who were very busy in harvesting their wheat fields. The wheat was very good. It was the first good harvest they had had for several years. Toward evening we passed a sand-draw leading from a small valley between the hills. We drove into this valley and camped there.

August 8. — In the forenoon we botanized on

the surrounding hills and in the sand-draw. The vegetation of the valley was not very rich. The valley was used as a pasture, and this fact, undoubtedly, was at least to a great extent the cause. I shall mention a few of the plants we collected: *Distichlis spicata* var. *stricta* [D. spicata], *Andropogon scoparius* [Schizachyrium scoparium], *Sporobolus cuspidatus* [Muhlenbergia cuspidata], and a variety of the same (No. 431) [Muhlenbergia cuspidata], *Eriogonum alatum, Urtica gracilis* [Urtica dioica subsp. gracilis], *Euphorbia cuphusperma* [E. davidii], *hexagona*, and *serpyllifolia, Cheylanthes lanuginosa* [Cheilanthes feei], *Amalanchier* [*sic*] *alnifolia, Prunus demissa* [P. virginiana] and *pumila* [P. pumila var. besseyi].

In the afternoon we took up our journey again, drove into the Big Horn Valley. From there we ascended the tableland between the Pumpkin Valley and the Lawrence Fork and arrived at Mr. E. Frostrom's in the evening. Here we stayed over Sunday.

August 9. — Sunday, no botanizing.

August 10. — We drove across the divide. This little divide of a width of 7 or 8 miles is, in my belief, the best strip of land in the whole region. The soil is rich. The bunch grass (*Carex* sp., No. 400 of my list) [C. filifolia] is very common, and this fact is regarded as a good sign on the upland prairie. Although the divide is higher than any other tableland in the region, the settlers have had a crop every year, while there has been none to speak of in the vicinity for 3 to 5 years.

We drove down into the Lawrence Fork, but turned to the west this time, and followed one of the branches upwards. Here we collected *Bidens chrysanthemoides* var. [B. cernua], *Geum strictum* [G. aleppicum], and *Ambrosia trifida* var. *integrifolia* [A. trifida]. We ascended the hills again. In a small hollow, we gathered *Coreopsis tinctoria*. When we had reached the tableland, we came to a settled region again. The fields seemed well tended to, and the crop was good.

We had now turned to the south-southwest. The further we drove, the more sandy, hilly, and rocky the land became. At noon, we stopped in a little valley surrounded by rocky hills, which were covered with very little soil. Here we botanized for a while, and found among others, *Bouteloua hirsuta, Sc[h]edon[n]ardus texanus* [S. paniculatus], *Lesquerella engelmannii* [Physaria ovalifolia], *Astralgalus missouriensis* [A. m. var. missouriensis] and *triphyllus* [A. hyalinus], *Oxytropis* sp. (No. 82) [O. multiceps], *Applopappus nuttallii* [Xanthisma grindelioides] and *spinulosus* [Xanthisma spinulosum], *Helianthus rigidus* [H. pauciflorus subsp. subrhomboideus], *Chaenactis douglasii* var. *alpina* (?) [C. douglasii], *Krynitzkia sericea* [Cryptantha cana], *Aphyllon ludovicianum* [Orobanche ludoviciana], *Eriogonum flavum, Euphorbia maculata* and *fendleri*.

In the afternoon, we drove several miles without seeing a house. We crossed a big sand-draw, a branch of the Lawrence Fork. When driving up the hills on the other side, Mr. Flodman stepped off and walked. While waiting on him I heard a wagon coming on the same road. Mr. Flodman had found a tufted, slightly woody plant, which I have identified as *Phlox muscoides* [P. bryoides]. When the wagon overtook us, we asked the man if he knew any place in the neighborhood where we could get water. He answered that we might get it at the next house about a mile away. We did so, but had to pay 25¢ for watering our horses that evening and the next day.

August 11. — In the forenoon we went down into the sand-draw mentioned above. We collected, among others, *Euphorbia maculata* and *fendleri*, *Oenothera pinnatifida* [Oe. albicaulis] and *albicaulis* var. (No. 112), *Mentzelia nuda, Phlox muscoides* [P. bryoides]. In the afternoon we took up our journey again, and arrived at Kimball a little before sunset.

August 12. — At first we botanized on the prairie around our tent. Here we gathered *Petalostemon tenuifolius* (No. 59) [Dalea purpurea var. purpurea], *Oenothera pinnatifida* [Oe. albicaulis] and *albicaulis* [nuttallii], *Veronica peregrina* var. (No. 283) [V. peregrina var. xalapensis], *Aphyllon ludovicianum* var. (No. 291) [Orobanche ludoviciana], *Eriogonum microthecum* var. *effusum* [E. effusum var. effusum], *Cnicus ochrocentrus* [Cirsium ochrocentrum], and *Oxybaphus angustifolius* [Mirabilis linearis]. We then went down to an irrigated meadow, extending along the Lodgepole Creek. The vegetation was luxurious; the grass reached over the knees. Here we found *Agropyrum* sp. (No. 467) [Elymus smithii?], *Ranunculus cymbalaria* and *hispidus* (No. 7 of my list) [R. macounii], *Nasturtium palustre* [Rorippa palustris], *Epilobium adenocaulon* [E. ciliatum], *Solidago canadensis, Aster commutatus* [A. falcatus subsp. commutatus?], *foliosus* [?], and *adscendens* (?) (No. 162) [?], *Eustoma russellianum* [E. grandiflorum], *Physalis longifolia,* Nutt. [P. hispida], *Orthocarpus luteus, Polygonum lapatifolium* [*sic*], and *Salix longifolia* [S. interior]. On the hills near it, we found small specimens of *Helianthus annuus*, scarcely distinguishable from *Helianthus petiolaris*. We then crossed the railroad and walked up to a sand-draw southwest of Kimball. Here we found among others, *Polanisia trachysperma* [P. dodecandra subsp. trachysperma], *Cleome integrifolia* [C. serrulata], *Acerates auriculata* [Asclepias engelmanniana], *Eriogonum cernuum* [E. effusum var. effusum], and *E. corymbosum* [E. nebraskense]. Returning over the prairie we picked some specimens of *Astragalus pubentissimus* (No. 74 of my list) [A. lotiflorus].

August 13. — In the morning, we took the trunk and the boxes to the railroad station and sent them by express to Lincoln. As the flora of the Lodgepole Valley has been worked up during the U. P. Railroad survey, we did not care to follow this valley. From the editor of one of the local papers, I heard that about 10 miles southeast of Kimball, there was a big draw, which runs down to Sidney. We therefore took a road running in a southeasterly direction from town. We came to a draw, where we stopped for dinner and botanized for a while, but did not find anything of interest. We then drove down the draw, knowing that it emptied into the Sidney Draw. The horses had not received any water since in the morning, which was hard on them, as we did not reach any house before sunset.

August 14. — The owner of the place where we stopped had a big herd of cattle but did very little farming. He had a very fine tree claim, which he had tended to very well. There will grow trees even on the dry plains of western Nebraska, if the planting and cultivating is done carefully, as was shown here. The trees were vigorous, although not very tall for their age. Nearly all had grown. They were, boxelder, green ash, Russian mulberry and catalpa. In the grove, we collected the following plants: *Cenchrus tribuloides* [C. longispinus], *Astragalus microlobus* [A. gracilis], *Euphorbia* sp. (No. 336) [E. effusum var. effusum], and *Sophora sericea* [S. nuttalliana]. From there we drove eastwards over the prairie and found a few acres of land covered by *Sophora sericea* [S. nuttalliana] in fruit.

About 6 miles from the place where we stopped overnight, we came to a settlement of Norwegians, who had broken a larger part of their homesteads, up to 120 acres. The average amount of open land the settlers of western Nebraska had, I believe, was not over 20 acres apiece. One of the Norwegians claimed that he had raised 40 bushels of barley to the acre this year. There will be good crops in western Nebraska whenever there is enough rain, which it has been this year. We drove about 40 rods out of the road to some hills bordering a draw. Here I walked around for a short while. I found *Prunus demissa* [P. virginiana] and *pumila* [P. pumila var. besseyi], *Ribes aureum* [R. odoratum] and *cereum* [R. c. var. inebrians], *Eriogonum corymbosum* [E. nebraskense], and *Franseria dumosa* [Ambrosia tomentosa], of which the latter two are not recorded for Nebraska. We then took the road leading to Potter, where we wished to be before night, as Mr. Flodman had to take the 8 o'clock train home. Before we reached Potter, we were caught in a heavy shower of rain.

August 15. — At Potter, the Lodgepole Creek has sunken into the sand. It is said that the water is reached first 20 or 25 feet below the surface. The channel is marked by a big sand-draw. In this sand-draw I botanized for several hours. The following were in the collection made: *Franseria hookeriana* [Ambrosia acanthicarpa], *Ambrosia psilostachya, artemisifolia,* and *trifida* var. *integrifolia, Artemisia canadensis* [A. campestris subsp. caudata], *biennis, ludoviciana,* and *filifolia, Croton texensis,* and *Portulacca* [*sic*] *oleracea.*

In the afternoon, I had full work with pressing plants and changing papers in the presses. In the evening I drove to Mr. Oscar Anderson's, a friend of mine who lives 4 miles north of Potter, and stayed over Sunday.

August 16. — Sunday, no botanizing.

August 17. — I went down the big draw that runs a mile or two from Mr. Anderson's. Here I spent the whole forenoon without finding anything new. I saw some plants I had collected before as for instance, *Eriogonum microthecum* [E. effusum var. effusum] and *Euphorbia fendleri.* Here and there, grew some woody plants, but only as bushes, as for instance, *Prunus demissa* [P. virginiana] and *pumila* [P. pumila var. besseyi], *Rosa fendleri* [R. woodsii], *Rhus toxicodendron* [Toxicodendron radicans var. rydbergii] and *canadensis* var. *trilobata* [R. aromatica].

In the afternoon, I started again, and drove 6 or 7 miles eastwards on the divide. As I saw that it was going to rain, I stopped a little earlier than usually, in order to get my things under the tent before the rain came. It rained hard the whole night. I had to dig a small ditch within the tent to protect my bed from water.

August 18. — When I had hitched on the horses and was ready to start, I saw that I had forgotten to fasten the water bucket. When I was trying to do so, the horses started to run. I could not run fast enough to catch the reins. The horses ran in a big circle, turned in on the road, followed this till they came to the house, ran around the water wheel and stopped against the fence of the corral. The owner of the place, who was occupied in filling the water trough, caught them there. I tied them to a post, and went around gathering up presses, satchels, collecting case, and tent, which were scattered over the prairie and the yard. When ready again, I drove down into the Lodgepole Valley, which I followed downward. At noon I stopped near the creek and botanized. I gathered, among others, *Typha latifolia, Sparganium eurycarpum, Glyceria nervata* [G. striata], *Leersia oryzoides, Andropogon furcatus* [A. gerardii subsp. gerardii] and *hallii* [A. gerardii subsp. hallii], *Sporobolus cryptandrus,* and *Gerardia tenuifolia* [Agalinis tenuifolia var. parviflora]. In the evening I arrived at Sidney, where I stopped overnight at a farmhouse about one mile from town.

August 19. — I went up to the creek and crossed it on a log that had been thrown across. In and near the creek I collected, among others, *Juncus nodosus,* var. (No. 381 of my list), *Catabrosa aquatica, Glyceria distans* [Puccinellia nuttalliana], *Andropogon scoparius* [Schizachyrium scoparium], *Muhlenbergia ramosa* [M. racemosa], *Sporobolus airoides, Ranunculus sceleratus, Epilobium adenocaulon* [E. ciliatum], *Euphorbia marginata,* and *Salix longifolia* [S. interior]. I then went up to the hills south of the valley. These hills had very rocky sides and flat tops covered with a very thin layer of earth. On the rocks, I found several kinds of lichens. On and around the hills, I collected *Aristida purpurea, Panicum capillare, Psoralea tenuifolia* [*sic,* for tenuiflora] [Psoralidium tenuiflorum], *Symphoricarpos occidentalis, Solidago nemoralis* var. (No. 158) [S. mollis], *Thelesperma filifolium, Eriogonum annuum* and *cernuum.* Of woody plants, I saw *Prunus demissa* [P. virginiana], *Celtis canadensis* [C. occidentalis], *Symphoricarpos, Ribes aureum* [R. odoratum] and *cereum.*

I then went down into a small draw where I found two species of *Physalis,* of which I have made some notes in my List (No. 271 and No. 273) [P. longifolia and P. hispida]. After having crossed a vast field of sunflowers, I came to the creek again about two miles above the farm houses. The creek was here running in the sand. A little further down it came to the surface again. Here I found a *Polygonum,* which I regard as a large, decumbent variety of *P. erectum* (See my list)[P. buxiforme].

After dinner, I took up my journey again, passing through Sidney. Sidney is a fine-looking town, depending to a great extent of the large number of trees that were planted there. Especially is this the case around the Fort.

Between Sidney and Lodgepole, I passed through one large pasture after the other, and saw the biggest herds of horses I ever saw, thousands, and again thousands. The Lodgepole Valley is here used exclusively for pastures and haylands. In the pastures, the grass was well kept down, and on the haylands it was newly harvested. I saw big stacks all over, where the land was not used as pastures. I arrived at the town of Lodgepole very late. After having watered my horses and picketed them out, I was going to put up my tent. I found that I had happened to stop on a knoll of sand. It was blowing hard, and the tent blew down twice. It took me about an hour, before I could make it stand.

August 20. — I botanized for a couple hours

around Lodgepole but did not find much of interest. The most noticeable were *Eliocharis* [*sic*] *acicularis*, *Ranunculus circinatus* [R. longirostris], and *Bidens chrysanthemoides* (?) [B. cernua].

After dinner I took up my journey again, and arrived at Chappell in the afternoon. I went into the Deuel County Bank and had a conversation with Mr. Hamilton. The horse that had been hurt while in our charge was better than they had ever expected. However it had become stifled, and the owner was to be paid some money for damage. Mr. Hamilton proposed to have the sum settled by arbitration, he selecting one man for the owner who was at Salt Lake City, and I the other. I drove over to Rev. Bergquist's before night.

August 21. — It had been raining during the night. It was cold and cloudy. Small showers of rain. I was tired out from the journey. I did not like to go out collecting in the bad weather but tended to the presses, which needed to be looked over, labeled and packed together some plants, etc.

August 22. – Rev. Bergquist and family had prepared themselves to drive out to McCuligan Canyon and pick berries. In the canyon, as stated before, there grows a great quantity of western cherries, currants, and grapes. The latter were not ripe yet but were large enough for jelly. I had intended to go out to the canyon some day before I left. I therefore made company, taking Rev. Bergquist's three boys in my buggy. It was early in the morning when we started, and late at night when we returned home. We were all satisfied with the day. Rev. Bergquist had several baskets full of berries, and I had made a collection of plants. The collection contained among others, *Andropogon scoparius* [Schizachyrium scoparium], *Clematis ligusticifolia*, *Dalea laxiflora* [D. enneandra] and *aurea*, *Mentzelia nuda*, *Liatris punctata*, *Artemisia dracunculoides* [A. dracunculus] and *filifolia*, *Polygonum dumetorum* [P. scandens], *Euphorbia petaloidea* [E. missurica var. petaloidea], *glyptosperma*, and *hexagona*.

August 23. — Sunday, no botanizing.

August 24. — Monday morning I started out again, and drove down to the draw Mr. Flodman and I visited on June 27. Near the springs in which we had collected *Limosella aquatica*, I stopped, picketed out the horses, and botanized in the vicinity. Here I found some plants of interest, as for instance *Crysopsis* [*sic*] *villosa* var. *sessiliflora* [Heterotheca villosa], *Aster commutatus* [A. falcatus subsp. commutatus?], *Pectis angustifolia*, *Artemisia canadensis* [A. campestris subsp. caudata], *Mollugo verticillata*, *Froelichia gracilis*, and a species of *Polygonum* (maybe P. *minimum* or *torreyi*, No. 345 of my list) [P. ramosissimum], of which the last two are new to Nebraska.

After dinner, I drove down the draw, which runs out into the South Platte Valley. In the draw, I collected *Asclepias arenaria* and *Petalostemon villosum* [Dalea villosa], both in fruit. When I came near the mouth of the draw, I found it fenced in for pastures. I therefore drove southwards across the hills which separate it from the Platte Valley. This was a hard task as the hills are very steep in several places. In two places the horses refused to go down, until I used the whip. I came down into the Platte Valley near the stopping place (I have forgotten the name) on the Union Pacific Rairoad between Julesburg and Big Springs. Here on the bottomland I collected *Cleomella angustifolia*. As it looked as if it was going to rain, and I had not taken the tent along, I had to see that I came to some house at night. I knew that I was about 5 miles from Julesburg, and that from this place to Big Springs it was 11 or 12 miles. I did not care to go to Julesburg, partly because it is in Colorado, and partly because I must come there before leaving the field. I thought that I could drive 6 or 7 miles easily before dark and started for Big Springs but had not taken the condition of the road in consideration. A few miles east of the place where I entered the valley, the Platte runs close to the bluffs. It has, therefore, been necessary to make the road over the bluffs. These are traversed by canyon after canyon; the road goes up and down, down and up. I had to drive in walk [*sic*]. In several places the rain had washed out the road, and the way had to be taken wherever possible. I had to use the whip a couple times to make the horses go. I afterwards wondered that the old buggy did not break. It became dark and began to thunder and rain, although not heavily. I saw a light at a distance. It seemed to be very near, and I began to hope, but it must have been 3 miles away, judging from the time it took to drive there. I did not see the road but relied on the horses. Suddenly they stopped. I leaned forward to see what was the matter. I found that the road was washed out. The rain had made a ditch 5 or 6 feet deep. The buggy stood with two wheels and one horse on each side of the ditch. I did not know at first what to do. I succeeded, at last, to get out of the dilemma by pulling hard on one rein and giving the horse on the other side a good lash by the whip, and turning the wagon at right angles, which ran on three wheels over the ditch, the swiftness and lightness of the buggy only hindering it from tipping over. When arriving at Big Springs, I put my horses in a livery stable and took supper and room at a hotel.

August 25. — In the morning I went down to the South Platte River, botanizing there for some hours. I found, among others, *Chrysopogon nutans* [Sorghastrum nutans], *Nasturtium obtusum* [Rorippa tenerrima], *Petalostemon violaceum* [Dalea purpurea var. purpurea], *Plantago lanceolata, Euphorbia glyptosperma, Vernonia fasciculata, Scirpus lacustris* [Schoenoplectus tabernaemontani], *Gerardia tenuifolia* [Agalinis tenuifolia var. parviflora], *Salix longifolia* [S. interior], etc.

Returning home, I took the road over the prairie, but did not find anything of interest. I arrived at Rev. Bergquist's house at 5 o'clock in the afternoon, and spent the rest of the day in looking over the presses.

August 26. — Rev. Bergquist had asked me to take part in a church festival to be held at Froid. At last I promised him to do so. According to this promise, I took half a day free from work in the afternoon. I went to Chappell to see that I could get settled the matter about the horse and also to return the wagon and one of the horses we had had. The other belonged to a neighbor of Rev. Bergquist. When down at Chappell, I found that Mr. Wolf, who had doctored the horse and still had it in his charge, had gone to Lodgepole and would not be back before the next day. We agreed to settle by arbitration. Mr. Hamilton had chosen Mr. Wolf as his man. I selected Mr. Swanson, a dealer in General Merchandise at Chappell.

August 27. — In the forenoon I botanized a little around home but did not find anything of interest. In the afternoon, Rev. Bergquist took me to Chappell. He, Mr. Hamilton, Mr. Swanson, and I rode out to Mr. Wolf's place. The horse was driven in from the pasture and examined. Mr. Swanson and Mr. Wolf settled upon the sum of $15 as the amount to be paid for damages. To this came a doctor bill of $15, which made in all $30, fully half the value of the horse. I could not spare that much of the money I had with me, but had to send a part after I had come home to Wahoo.

August 28. — In the forenoon I packed my things, looked over the presses, boxed up plants, etc. In the afternoon, Rev. Bergquist took me to Julesburg. Here I sent a box of plants and two presses by express to Lincoln. Mr. Flodman had taken one press; the other two, I intended to take with me into the cars. As it was still a few hours before the train left, I took a walk down to the Platte Valley bottomland. I collected, among others, *Applopappus rubiginosus* (No. 145 of my list) [Rayjacksonia annua], *Bidens cernua* var. (No. 190), and *Cleomella angustifolia*. As there was no connection at Sterling where the Union Pacific Railroad crosses the Chicago, Burlington and Quincy Railroad, and it had taken me two days to come to Lincoln, if taking that route, I took the Union Pacific Railroad to Kearney City, where the connection was better. I arrived at Kearney at 4 o'clock in the morning of [*continues below:*]

August 29. – As the train for Lincoln left at half past eight, I took a walk down to the Platte, where I botanized on the islands. I collected, among others, *Petalostmon candidus* [Dalea candida var. candida], *Amorpha fruticosa, Millelotus* [*sic*] *alba* [Melilotus alba], *Helianthus maximilianii, Gerardia purpurea* [Agalinis purpurea], and *Desmodium canadense.*

I returned to town, carrying my satchels and presses to the Chicago, Burlington and Quincy depot, and bought a ticket. The baggage man was very busy. When I asked him to check my satchels, he asked me to wait a little until he had checked some trunks. He held 6 or 7 checks in his hand. He would be back very soon. When the train was ready to pull out, I went out. Then he told me to take the satchels into the cars. As I did not like to leave my presses and collecting case, which were in the waiting room, I preferred to stay to another train. Hearing that there was no other train on that branch before night, I gave the baggage man a scolding, got the money for my ticket back, and took a livery team to Axtell on the main road. I stopped here and ther[e] on the way, collecting. On the Platte bottom, I found, among others, *Dalea alopecuroides* [D. leporina] and *Desmanthus brachylobus* [D. illinoensis]. I then crossed the sandhills, which I have described before. Here I collected *Petalostemon villosum* [Dalea villosa], *Gilia longiflora* [Ipomopsis longiflora], *Froelichia floridana*, etc. After my arrival at Rev. Thoren's I decided to stay over Sunday there, as the evening train came very late at night to Lincoln, and the morning train left early Monday morning. I was out early and got a supply of *Franseria tomentosa* (No. 181 of my list) [Ambrosia grayi], which had not been recorded for Nebraska, although it already has become a troublesome weed in Kearney Co., especially on low, rich soil. It is, as the quackgrass, hard to exterminate on account of its creeping rootstock.

August 30. — Sunday, no botanizing.

August 31. — Early in the morning I left for Lincoln, where I began to classify my collection.

September 1–15. — I worked in the Herbarium of the University of Nebraska, classifying my collection, writing labels, preparing my list of plants, and packing and sending nearly half of the collection. The remainder I have sent as soon as ready.

† † †

SUMMARY

I shall here add a short summary of my observations in western Nebraska Deuel [including present-day Garden], Cheyenne [including present day Morrill], Banner, Scotts Bluff, and Kimball Counties.

The land in this region consists mostly of high tableland rising from about 3000 feet over the sea in southeast Deuel Co. to over 5000 feet in Scotts Bluff Co. This tableland is traversed by valleys, from 100 to 400 feet lower. The most important of these valleys are:

North Platte Valley, running northwest and southeast through Scotts Bluff, Cheyenne [present-day Morrill], and Deuel [present-day Garden] Counties.

South Platte Valley, which crosses the southeast corner of Deuel Co.

Lodgepole Valley, running west and east through Kimball, Cheyenne, and Deuel Counties.

Pumpkin Valley, running west and east through Banner Co., and a branch of the latter, the Lawrence Fork, extending into Kimball Co.

The descent from the tableland into the valleys is very steep and rocky. Canyons, some 200 to 300 feet deep and only a few rods wide, run from the valleys deep into the tableland. The divide between the North Platte Valley in Scotts Bluff Co., and the Pumpkin Valley is higher and very narrow, and takes more the character of a ridge of hills than that of a tableland. The ridge dividing the Platte Valley in said county into the Cedar and Mitchell Valleys is a branch of this divide.

Climate

Western Nebraska has suffered much from drought in several years, causing a total failure of crops. This year, there has been enough rain, however, and the crop has been good. The winters are said not to be very severe, and there is little snow. The cattle feed out throughout the winter, and have to be given hay only in bad weather. The summers are dry, especially so the latter part. The western portion is frequently visited by hail storms, which do considerable damage.

Drainage

The tableland is traversed by "sand-draws" or "wadies" that run through canyons or gorges down into the valleys. In the canyons, are springs not uncommon, which either sink down into the sand or give rise to creeks. The creeks of western Nebraska are all small except Lodgepole Creek and Horse Creek, which both come from Wyoming. Among the creeks may be mentioned: Rush Creek and Cedar Creek in Cheyenne Co., Pumpkin Creek in Banner Co., and Kiowa Creek in Scotts Bluff Co.

Soil

On the tableland the soil is generally rich, although a little sandy. On the bluffs, where the tableland stoops into the valleys, the layer of soil is very thin, and the rocks generally come up to the surface. The bottomland along the streams is rich from deposits, but sometimes alkaline. The rest of the valleys, or the so-called "second bottom," is sandy and far inferior to the tableland. North of the North Platte River in Scotts Bluff Co., above the irrigation ditches, I saw among the poorest land I have seen.

Vegetation

In a dry climate, as it is in western Nebraska, the vegetation is naturally low. Along the streams it is more luxurious. I could distinguish five different floras, viz:

1. The Upland Prairie Flora
2. The Cliff and Hill Flora
3. The Canyon Flora
4. The Sand-draw Flora
5. The Bottomland Flora

The characteristic plants of the first are the upland grasses: grama grass, *Bouteloua oligostachya* [B. gracilis], buffalo grass, *Buchloë dactyloides*, bunchgrass, *Carex* sp. (No. 400 of my List) [C. filifolia, NEB 192032, Deuel Co., June 25, in Rydberg's hand; possibly NY 2275873, with the identical printed label but with no.1078 and all other details not in Rydberg's hand], the several kinds of cactuses and thistles.

The plants, characteristic for the second, are low, often tufted, perennials, as: *Eriogonum flavum* and *multiceps* [E. pauciflorum var. gnaphalodes], *Musenium* [*sic*] *tenuifolium*, *Arenaria franklinii* var. [Eremogone hookeri], *Gilia pungens* var. *caespitosa* [Leptodactylon caespitosum], *G. iberidifolia* [Ipomopsis congesta], *Phlox hoodii*, *Astragalus caespitosus* [A. spatulatus], *triphyllus* [A. hyalinus], and *sericoleucus*, etc.

The plants of the canyons were of a greater variation. To this flora belong most of the woody plants of this region. Among herbaceous plants, I shall mention a few: *Clematis ligusticifolia*, *Mentzelia ornata* [M. decapetala], *Brickelia [sic] grandiflora*, *Oenothera caespitosa*, etc.

The sand-draw, however furnishes the best display of flowers. Among plants growing in or near the sand draws, I will mention: *Argemone platyceras* [A. polyanthemos], *Mentzelia nuda*, *Asclepias speciosa*, *Acerates auriculata* [Asclepias engelmanniana], *Polanisia trachysperma* [P. dodecandra subsp. trachysperma], *Cleome integrifolia* [C. serrulata], *Oenothera albicaulis* [Oe. nuttallii], *coronopifolia*, and *pinnatifida* [Oe. albicaulis], *Crysopsis* [*sic*] *villosa* [Heterotheca villosa], etc.

The bottomland flora consisted mostly of grasses and members of the order Cyperaceae and a few woody plants.

Of course, in some places, these five floras mixed together as, for instance in the Lawrence Fork, which is a narrow valley where the flora was a combination of the first three. Otherwise, most of the plants were confined to their peculiar locality.

Native Wood

The woody plants of western Nebraska are mostly confined to the canyons and the valleys near the water courses. There are nowhere any forests, only small groves or scattered trees. The trees are of middle size or, oftener, below. The largest tree I saw was a red ash, *Fraxinus pubescens* [F. pennsylvanica], near the Hackberry Springs. Of deciduous trees I give the following list. western cherry, *Prunus demissa* [P. virginiana], plum, *Prunus americana*, cottonwood, *Populus angulata* [P. deltoides var. monilifera], and another poplar near *P. balsamifera* and *P. angustifolia* (in Carter Canyon, No. 372 of my list) [P. ×acuminata], hackberry, *Celtis occidentalis*, white elm, *Ulmus americana*, green and red ash, *Fraxinus viridis* and *pubescens* [two forms of F. pennsylvanica], boxelder, *Negundo aceroides* [Acer negundo], willow, *Salix cordata* var. *vestita* [S. eriocephala] and several varieties of *S. longifolia* [S. interior]. Of shrubs I noticed: wild currants, *Ribes aureum* [R. odoratum], *cereum*, and *floridum* [R. americanum](canyon in Kiowa Valley), sand cherry, *Prunus pumila* [P. pumila var. besseyi], buffalo berry, *Shepherdia argentea* (islands in Platte), poison ivy, *Rhus toxicodendron* [Toxicodendron radicans var. rydbergii], *Rhus canadensis* var. *trilobata* [R. aromatica], *Rosa blanda* [R. woodsii], *fendleri* [R. woodsii], *woodsii*, and *nutkana* [R. woodsii] (McCuligan Canyon), *Symphoricarpos occidentalis*, *Bigelovia graveolens* var. *glabrata* [Ericameria nauseosa var. graveolens], *Atriplex nuttallii* [A. canescens], *Amorpha fruticosa*, (Big Springs) and *canescens*, *Artemisia cana* and *filifolia*, *Eurotia lanata* [Krascheninnikovia lanata]. To these are to be added two woody climbers, wild grapes, *Vitis riparia*, and Virginia creeper, *Ampelopsis quinquefolia* [Parthenocissus vitacea; *P. quinquefolia* is not in range].

Native Grasses

Two grasses and one sedge, the grama, the buffalo grass and the niggerhead or bunchgrass, as stated before, constitute the larger part of the flora of the Upland Prairie. How many millions of cattle, horses, and sheep have sustained their lives, both summer and winter, on these low grasses.

The bottomlands, especially near the Platte River and Lodgepole Creek, are covered mostly by grass and used extensively for pastures and haylands. The most valued grasses for haymaking are *Agropyrum glaucum* [Elymus smithii], *Panicum virgatum*, *Andropogon scoparius* [Schizachyrium scoparium], *Bouteloua oligostachya* [B. gracilis], *Elymus canadensis*, *Muhlenbergia ramosa* [M. racemosa], and a *Poa* [Poa arida].

Weeds

The worst weed of western Nebraska, I believe, is *Chenopodium leptophyllum* [C. pratericola] and, next to it, the sunflowers, *Helianthus annuus* and *petiolaris*. I found these everywhere on broken land as well as on the prairie, but most common in old fields. More local were *Chenopodium album* [C. berlandieri], *Solanum rostratum*, *Amarantus [sic] albus* and *blitoides*. *Ambrosia trifida* and *artemisiifolia*, troublesome weeds in eastern Nebraska, seem not to have acquired a foothold yet in the west. The only place where I found them in any number was in the sand-draw of Lodgepole, near Potter, where they grew together with two related plants, *Ambrosia psilostachya* and *Franseria hookeriana* [Ambrosia acanthicarpa], and, of course, did not do any harm there. The sandburs, *Cenchrus tribuloides* [C. longispinus], I found only at one place, in a tree plantation south of Dix, Cheyenne Co., and the purslane, *Portulaca oleracea*, on the railroad bank near Potter. *Solanum triflorum* and *Krynitzkia crassisepala* [Cryptantha minima], which grow profusely where the ground has been disturbed in some way, as for instance in the prairie dog towns, on account of this habit, might become troublesome weeds. At last I will add the names of two plants, not recorded before in Nebraska, to the blacklist of weeds. Both are said to be troublesome in the locality in which they grow, namely *Franseria tomentosa* [Ambrosia grayi] in Kearney Co., [No. 181 of the *List*] and *Amarantus* [*sic*] sp. [A. blitoides] in Pleasant Valley, Scotts Bluff Co. (No. 313 of my list).

Prospects

Western Nebraska would become a good farming country if it received a little more rain than it has and if the rainfall was more distributed throughout the season. Maybe a change will take place as it has in eastern Nebraska, where the rainfall has increased considerably since the land became broken. This year there has been more rain than usually, and the settlers have had a good crop. Irrigation has been much talked about, and plans have been laid to irrigate the whole tableland of Deuel Co., which however was an impossibility, on account of the altitude of that land. In Scotts Bluff Co., there are irrigation ditches in operation, and still more are under building. Under irrigation are raised fine crops.

† † †

Luther Academy, Wahoo, Neb., February 8, 1892

Rydberg's List of Plants from the 1891 Trip, with Notes About the 1890 Trip

The following text is verbatim from Rydberg's manuscript, except for modernized spelling of place names, simplified punctuation, and all bracketed matter, including citation of herbarium specimens. Bracketed names are updated or corrected nomenclature, mostly following *Flora of Nebraska,* second edition, 2011. The present authors' notes and comments are in brackets within the species descriptions and in reduced font following them. Holzinger's comments about the specimens are also verbatim; they are here inserted after each species' description, but in Rydberg's original *List* they are given as one unit. The sites are shown on the maps in Figures 1 and 2 and the photographs in Figures 7–19.

LIST of PLANTS collected by P. A. Rydberg and Julius Flodman during the summer of 1891, mostly in western Nebraska

[CHARACEAE]

[*Chara foetida* A. Br., *det.* F. K. Daily in 1943; July 16, 1891, *s.n.*, Banner and Kimball counties; NEB 227432]

Rydberg mentioned these collections in the *Report* but not in his *List*. The specimens from both sites are mounted on the same sheet, NEB 227432.

RANUNCULACEAE

1. ***Clematis ligusticifolia*** Nutt. – Climbing over bushes; common in western Nebraska, rare in the eastern* part of the state. It was collected by me in 1889, in Saunders Co. Canyon, south of Pumpkin Seed Valley, July 10 [NEB 173367]; McCuligan Canyon, Deuel [now Garden] Co., Aug. 22.

Not cited in the manuscript is NEB 173391: Kiowa Valley, Scotts Bluff Co., July 28, 1891. Rydberg also collected it in Banner Co. in August, 1890 (NEB 173395, 207912). In 1890, he identified his 1890 specimen (NEB 173395) from that county as *C. virginiana*, and in 1926 he annotated it as *C. missouriensis* Rydb. (now a synonym of *C. virginiana*), but it is closer to *C. ligusticifolia* by its leaves that are densely pubescent beneath.

**Clematis virginiana* L. is only in the eastern half of Nebraska, including Saunders Co., and *C. ligusticifolia* is only in the western half and not in or near Saunders County. The ranges of the two species abut only in the northernmost Nebraska counties.

2. ***Anemone cylindrica*** Gray – Prairies, throughout the state. Platte River bottoms in Kearney Co., June 15 [NEB 173267].

3. ***Thalictrum purpurascens*** L. [*T. dasycarpum* Fisch. & Avé-Lall.] – Near water. Platte bottoms, Kearney Co., June 15 [NEB 173788]; Little Medicine Creek, Frontier Co., June 22 [NEB 173807].

4. ***Ranunculus circinatus*** Sibth. [*R. longirostris* Godr.] – In water. – It is not contained in Mr. Herbert J. Webber's CATALOGUE OF THE FLORA OF NEBRASKA. In Lodgepole Creek, near Lodgepole, Cheyenne Co., Aug. 29* ["Aug. 19" on label of NEB 173315].

*On August 29, Rydberg was in Kearney City, Buffalo Co., 180 miles east of Lodgepole; August 19 is therefore the correct date of collection.

5. ***Ranunculus cymbalaria*** Pursh – In springs and creeks in western Nebraska. Pumpkin Seed Creek, Banner Co., July 17 [NEB 173648]; spring in Pleasant Valley, Scotts Bluff Co., July 28; Horse Creek, Aug. 1

Rydberg also collected it in Banner Co. in August, 1890 (NEB 173647, 207913).

6. ***Ranunculus sceleratus*** L. – In water, rare. Horse Creek, Scotts Bluff Co., Aug. 1; Lodgepole Creek, near Sidney, Cheyenne Co., Aug. 19 [NEB 173724].

7. ***Ranunculus hispidus*** Hook. [*R. macounii* Britton] – New to the Region; and perhaps new to the United States. In the revised edition of Gray's Manual, the locality is given thus: "On the northern shore of Lake Superior, and north and westward; probably in N. Minn." Only a few specimens were collected on an irrigated meadow north of Kimball, Kimball Co., Aug. 12 [NEB 173710]. When collecting it, I took it for a form of *R. repens* or of *R. pensylvanicus*, luxuriant because of the irrigation.

8. ***Delphinium azureum*** Michx. – var.. . . . . . . *sic* [*D. virescens* Nutt.] – It differs from the type in being lower, in having the stem glandular hairy up to the raceme and the leaves often with broader segments. On the hills north of Lawrence Fork, Banner Co., July 9 [NEB 173451].

PAPAVERACEAE

9. ***Argemone platyceras*** Link. [*A. polyanthemos* (Fedde) G. Ownbey] –Sandy soil; common throughout Region.–Sand-draw in Deuel [or possibly Garden] Co., June 28.

Not given in Rydberg's *Report* or his *List* is NEB 173912, July 8, 1891, Lawrence Fork, Banner County. He also collected it in Deuel or Garden Co. in July, 1890 (NEB 173908, the specimen later so-annotated by him in October, 1926). He mentioned it often in his *Report*, and it is common in central and western Nebraska.

FUMARIACEAE

10. ***Corydalis curvisiliqua*** Eng. [subsp. *occidentalis* (Engelm. ex A. Gray) W. A. Weber] – I collected a few specimens of what I regard as belonging to this species among bushes in Lawrence Fork, July 8 [NEB 173939]. The outer petals are wing-crested, the seeds are shining to the naked eye, but finely muricate under the lens.

CRUCIFERAE

11. ***Draba carolin[ian]a*** Walt. var. ***micrantha*** Gray [*D. reptans* (Lam.) Fernald] – In fruit only; on high prairie land in Deuel Co., June 27 [NEB 174390].

12. ***Arabis holboellii*** Hornem. [*Boechera grahamii* (Lehm.) Windham & Al-Shehbaz] – A few specimens collected in Pumpkin Seed Valley, Banner Co., July 14 [NEB 174033], and one in Kiowa Valley, July 27 [or morning of the 28th and cited as *Draba holboelii*].

13. ***Arabis*** sp. . . . . *sic* [*Thelypodium integrifolium* (Nutt.) Endl.] – General appearance somewhat like *A. laevigata*, smooth and glaucous, branched above; root fusiform; stem-leaves lanceolate, entire, glabrous, not sagittate at the base; petals white, spatulate, twice as long as the whitish sepals; pods and root-leaves not seen. Low meadow on the upper Pumpkin Seed Valley, July 17 [NEB 174644].

J. M. Holzinger: "*Thelypodium integrifolium*."

14. ***Sinapis arvensis*** L. – Throughout the state.–Railroad bank, Kearney Co., June 15.

J. M. Holzinger: "*Brassica sinapistrum*."

15. ***Erysimum cheirant[h]oides*** L. – Railroad bank, Kearney Co., June 15

Rydberg did not mention this in his *Report*.

16. ***Erysimum asperum*** DC. – Hills of western Nebraska.–Deuel Co., June 25; Rush Creek [Garden Co.], July 2; south of Scotts Bluff, July 23.

Not mentioned in the *Report* is NEB 174163: Hills north of Kimball, Kimball Co., August 11, 1891. Rydberg also collected it in Deuel or Garden Co. in July, 1890 (NEB 174179).

17. ***Sisymbrium canescens*** Nutt. var. . . . . *sic* [*Descurainia pinnata* (Walter) Britton] – Pubescence short and glandular, as in *S.*

incisum Eng., but the seeds are in two rows. On the top of Wild Cat Mountains [Banner Co.], between 5,000 and 6,000 feet above the sea-level. July 16 [NEB 174569].

J. M. Holzinger: "*Sisymbrium incisum.*"

18. ***Nasturtium obtusum*** Nutt. [*Rorippa tenerrima* Greene] – Not in Mr. Webber's CATALOGUE. Unusually low specimens, collected on the sands of South Platte River, near Big Springs, Deuel County, August 25 [NEB 174451].

19. ***Nasturtium palustre*** Nutt. [*Rorippa palustris* (L.) Besser] – In water.–Platte River, Kearney Co., June 15 [NEB 174470]; Horse Creek, Scotts Bluff Co., Aug. 1; Lodgepole [Creek], near Kimball, Kimball Co., Aug. 12.

20 ***Nasturtium curvisiliqua*** Nutt. [*Rorippa sinuata* (Nutt.) A. S. Hitchc.] – I would have taken it for *N. sinuatum* if it had not been like the specimens of *N. curvisiliqua* Nutt. from Colorado in the Herbarium [NEB] of the University. It is new to Nebraska. A few specimens were collected near a well 1.5 miles from the Wyoming line, July 31 [NEB 174487].

On July 31, 1891, Rydberg was in western Scotts Bluff Co., near the Wyoming border.

21. ***Nasturtium sinuatum*** Nutt. [*Rorippa sinuata* (Nutt.) A. S. Hitchc.] – Not in Mr. Webber's CATALOGUE. Platte River, Kearney Co., June 15 [NEB 174417, 174428].

22. ***Lesquerella engelmannii*** (Gray) Wats. [*L. ovalifolia* Rydb., *Physaria ovalifolia* (Rydb.) O'Kane & Al-Shehbaz] – New to Nebraska. It was first collected by Mr. J. Flodman, who took it for a small form of *Physaria*, on the hills of the Upper Lawrence Fork, Kimball Co., Aug. 11 [NEB 77288, **isotype** of *L. ovalifolia* Rydb.; NY 185553 **holotype**; NY 185554 **isotype**; US 4849 **isotype**]

23. ***Lesquerella ludoviciana*** DC. [*Physaria ludoviciana* (Nutt.) O'Kane & Al-Shehbaz] – It is not uncommon on the hills of western Nebraska, and usually branched from the root. Deuel Co., June 25 [NEB 77271, 77272]; Banner Co., July 9; Scotts Bluff Co., Aug. 24.

Rydberg also collected it in Deuel Co. in July, 1890 (NEB 77269).

24. ***Physaria didymocarpa*** Gray [*Physaria brassicoides* Rydb.] – Not in Mr. Webber's CATALOGUE. Collected only in fruit in canyons south of Scotts Bluff. The specimens differ considerably from those in the Herbarium [NEB] of the University of Nebraska, being of twice the size or more, the cells of the pod not being spherical, but obovate, diverging at the top etc. July 22 [NY 185608 **holotype** of *P. brassicoides* Rydb.; **isotypes** NEB 77497, 174378; RM 2615; US 4739]

Rydberg called this *Physaria didyma* in his *Report.*

CAPPARIDACEAE

25. ***Polanisia trachysperma*** T. & G. [*P. dodecandra* (L.) DC. subsp. *trachysperma* (Torr. & A. Gray) Iltis] – Common in sand-draws throughout the region. Deuel Co., June 27, Aug. 24; Banner Co., July 8 [NEB 174733]; Kimball Co., Aug. 12.

Not cited in the *List* is NEB 174737: Rush Creek, Garden Co., July 2, 1891.
Rydberg collected this species in Deuel or Garden Co. in 1890: July (NEB 174741) and August (NEB 207888).

26. ***Cleome integrifolia*** T. & G. [*Cleome serrulata* Pursh] – Common in western Nebraska. Upper Pumpkin Seed Valley, Banner Co., July 17 [NEB 174656]; Kimball, Aug. 13.

27. ***Cleome angustifolia*** Torr. [*Cleomella angustifolia* Torr.] – South Platte bottoms, near Julesburg [Sedgwick Co.], Colo., August 24 [NEB 174695, 174697, 174705].

Given as "*Cleomella*" in the *Report.* In 1890, Rydberg collected this species in Deuel Co., near Lodgepole Creek (NEB 174698).

VIOLACEAE

28. ***Viola nuttallii*** Pursh–In fruit only, on the hillsides among stones. – Rush Creek, Deuel [now Garden] Co., July 2; Scotts Bluff, July 20 [NEB 180291].

POLYGALACEAE

29. ***Polygala alba*** Nutt. – Sandhills of Phelps Co., June 18; hills of McCuligan Canyon, Deuel [now Garden] Co., August 22; Scotts Bluff, July 22.

[SANTALACEAE, collected in 1890 but not 1891]

[***Comandra umbellata*** (L.) Nutt. subsp. ***pallida*** (A. DC.) Piehl – Lawrence Fork, Banner Co. July, 1890, NEB 171057]

This plant was not cited for 1891 in the *Report* or *List*. As is typical for his 1890 collections, the specimen does not have a Rydberg number on its label.

CARYOPHYLLACEAE

30. ***Arenaria franklinii*** Dougl. var. ***minor*** Hook. & Arn. [*Arenaria hookeri* Nutt. ex Torr. & A. Gray, *Eremogone hookeri* (Nutt. ex Torr. & A. Gray) W. A. Weber] – Hills south of Pumpkin Seed Valley, July 6 & 14.

J. M. Holzinger: "I consider this the species, *Arenaria franklinii*."

31. ***Arenaria franklinii*** Dougl. var. . . . . *sic* [*A. hookeri* Nutt. ex Torr. & A. Gray, *Eremogone hookeri* (Nutt. ex Torr. & A. Gray) W. A. Weber] –Both *A. Franklinii* and *A. pungens* are described as having the [petals] about equaling the calyx. In the specimens collected on Court House Rock, Cheyenne [now Morrill] Co., July 4, and on the hills of Scotts Bluff, July 23, the petals are nearly half longer.

J. M. Holzinger: "I consider this the species, *Arenaria franklinii*."

Julius Flodman collected this species on Court House Rock, July 4, 1891 (NEB 172843).

32. ***Arenaria*** sp. . . . . *sic* [*A. hookeri* Nutt. ex Torr. & A. Gray, *Eremogone hookeri* (Nutt. ex Torr. & A. Gray) W. A. Weber] – Very low and cushioned, 1–2 inches high; cyme 1 inch, few-flowered, generally 1-flowered; leaves about 3 lines long. It is perhaps a dwarfed variety of *A. franklinii*. Hills of the Lawrence Fork, Banner Co., July 8 [NEB 172838].

J. M. Holzinger: "I take it to be the var. *minor* Hook. & Arn., of this species *A. franklinii*."

Rydberg also collected this species in Banner Co. in August, 1890 (NEB 172852, 172854).

PORTULACACEAE

Portulacca *sic* ***oleracea*** Eng. – [Seen on the bank of Union Pacific Railway at Potter, Cheyenne Co., August 15, according to the *Report*, but not collected or cited in this *List*.]

33. ***Talinum teretifolium*** Pursh [*Phemeranthus parviflorus* (Nutt.) Kiger] – Sand-hills of Kearney Co., June 13 [RM 250536].

MALVACEAE

34. ***Callirhoe involucrata*** Gray – On sandy soil in Kearney Co., June 13 [NEB 179964].

35. ***Callirhoe alcaeoides*** Gray – Together with the preceding.

36. ***Malvastrum coccineum*** Gray [*Sphaeralcea coccinea* (Nutt.) Rydb.] –Hills and prairies; Deuel Co., June 25; Scotts Bluff, July 21 [NEB 180102]; Potter, August 14.

LINACEAE

37. ***Linum rigidum*** Pursh [*L. compactum* A. Nels.] – Hills, Phelps Co., June 18; Kearney Co., June 15 [NEB 179179]; Lawrence Fork, Banner Co., July 9.

> The June 15 and 18 collections are not cited in the *Report*, but June 25 is given for Deuel Co., south of Rev. Bergquist's home. Rydberg also collected it in Deuel Co. in July, 1890 (NEB 179180).

CELASTRACEAE

38. ***Celastrus scandens*** L. – Islands in Platte River, Kearney Co., June 15 [NEB 179507].

VITACEAE

[*s.n.*] ***Vitis riparia*** Michx. – Throughout Nebraska; in the west in canyons.

> The locations of Rydberg's specimens, if any, are unknown. This is the only species of *Vitis* that grows in the trip's range.

39. ***Ampelopsis quinquefolia*** Michx. [*Parthenocissus vitacea* (Knerr) A. Hitchc.] – as preceding [i.e., *Vitis riparia* above]. Lawrence Fork, July 9 [NEB 179810, 179827].

> This plant, woodbine, is cited as Virginia creeper in the *Report* for this date and for June 26, July 9, and July 28. Virginia creeper, as understood today, is *Parthenocissus quinquefolia* and does not occur this far west in Nebraska. *Parthenocissus vitacea* was recognized as a separate species in 1894, after Rydberg's trip, and it is the only species of *Parthenocissus* in western Nebraska.

SAPINDACEAE

40. ***Negundo aceroides*** Moench [*Acer negundo* L.] – Throughout Nebraska; in the western part only in canyons, near springs. Hackberry Springs, Banner Co., July 11.

ANACARDIACEAE

41. ***Rhus toxicodendron*** L. [*Toxicodendron radicans* (L.) O. Kuntze var. *rydbergii* (Small ex Rydb.) Erskine] – On the sides of the hills and in canyons. Kearney Co., June 15 [NEB 179110].

> Flodman and Hapeman were stricken with dermatitis from poison-ivy on June 13.

42. ***Rhus canadensis*** Marsh. var. ***trilobata*** Gray [*Rhus aromatica* Aiton] –Hills and canyons throughout the West. McCuligan Canyon, Deuel [now Garden] Co., June 26 [NEB 179148].

> In 1890, Rydberg collected this from Deuel or Garden Co. (NEB 207894).

LEGUMINOSAE

43. ***Thermopsis rhombifolia*** Nutt. – Sand-draws. In fruit, Rush Creek, Deuel [now Garden] Co., July 2 [NEB 91413].

44. ***Sophora sericea*** Nutt. [*S. nuttalliana* B. L. Turner] – Prairie and broken land.–Deuel Co., June 25; South of Dix, Cheyenne [Kimball*] Co. August 14 [NEB 177659, 177662, 177668].

> *Dix is in Kimball Co., near the Cheyenne Co. border.

45. ***Lupinus argenteus*** Pursh. var. ***decumbens*** Wats. – Hillsides and canyons. Lawrence Fork [Banner Co.], July 10 [NEB 177381, 177388]; Hackberry Springs [Banner Co.], July 11 [NEB 177379]; Cedar Valley [Scotts Bluff Co.], July 18 [NEB 177380].

> Nebraska's plants are assignable to var. *argophyllus* (A. Gray) S. Wats. and to var. *argenteus*, but some are intermediate. Rydberg also collected this species at Hackberry Springs in August, 1890 (NEB 177391, 207899).

46. ***Lupinus pusillus*** Pursh – Hills, Frontier Co., near Curtis, June 23 [NEB 177487]; Deuel Co. June 25 [NEB 311890].

47. ***Hosackia purshiana*** Benth. [*Lotus purshianus* (Benth.) Clements & Clements ex Otley; *L. unifoliatus* (Hook.) Benth.; *Acmispon americanus* (Nutt.) Rydb.] – Dooryard, Phelps Co., June 18; sand-draw, McCuligan Canyon, Deuel [now Garden] Co., August 22; Platte River bottoms, Kearney Co., August 29 [NEB 177020, 177025].

48. ***Psoralea tenuiflora*** Pursh [*Psoralidium tenuiflorum* (Pursh) Rydb.] – I prepared, last year, some notes on the *Psoraleas* of Nebraska, which I submitted to Dr. Britton of Columbia College, New York, together with specimens. From the material in the herbarium [NEB] of the University of Nebraska and in my own collections, I believed that we had two distinct species (or, at least, two well-marked varieties) under the name *Ps. tenuiflora* Pursh (*Ps. floribunda* Nutt.), one from eastern Nebraska, which I called *Ps. floribunda*, one from western Nebraska, for which I reserved the name *Ps. tenuiflora* Pursh. I made the following distinction:

Ps. floribunda: racemes dense, 3–5 in.; stem canescent, under the lens hirsute; leaflets generally 5, oblong, 1 to 1.5 inch; glands on the lower side of the leaves and one the calyx hidden by the hairs; teeth fo the calyx lance-ovate, the lower longer; flowers nearly 3 lines; standard orbicular, notched.

[***Ps. tenuiflora,*** mistakenly written as *Ps. floribunda* repeated from above]: racemes few-flowered, only 1–3 in.; bracts often with a single flower; stem minutely canescent; leaflets 3, oblong-lanceolate, .5–.75 in., planely [*sic*] punctate; calyx punctate, colored, teeth equal; flowers 2 lines; standard obovate.

Dr. Britton remarks that he had concluded that *Ps. floribunda* could not go unnoticed into *Ps. tenuiflora*, but recommended to put another season's field work on it, which I have done. The conclusion to which I have come is that they belong to one species. All specimens seen in the west fit my description of *Ps. tenuiflora*, except that the leaflets and the racemes are a little longer on more low and fertile soil. Specimens collected on the prairies of Phelps Co. (48b) are intermediate between the two, having some of the leaves 5-foliolate, longer, the racemes longer, and more hairy leaves and calyx; but the general appearance of them differs more from our eastern variety than from our western. Our eastern, more robust, more silky plant, of which I have no specimens preserved (there are several good specimens in the herbarium [NEB] of the University), I still believe has a good right to be separated as a well-marked variety. As the western form seems to have been the type of Mr. Pursh's description of *Ps. tenuiflora* and the eastern form the type of *Ps. floribunda* Nutt., an appropriate name for the latter would be *Psoralea tenuiflora* Push. var. *floribunda* (Nutt.).

On prairies: Deuel Co., June 27; Rush Creek [Garden Co.], July 2; Sidney, Aug. 19; (48b) Phelps Co. June 18 [NEB 178104].

J. M. Holzinger: "These *Psoraleas* gave me as much trouble as they did you. And I recognize the difference in the size of flowers and leaves of the several forms of your *P. tenuiflora*. If you do separate one of the extreme forms as *P. floribunda*, you will have plants in your own collection which may with equal right stand with the species or the variety. So I preferred not to do anything in the matter myself; I let it stand as *tenuiflora*."

Rydberg several times spelled this as 'tenuifolia' in the *Report*. He also collected it in Deuel or Garden Co., July, 1890 [NEB 171803].

49. ***Psoralea argophylla*** Pursh [*Pediomelum argophyllum* (Pursh) Grimes] – Prairies; Pumpkin Seed Valley [Banner or Morrill county], July 18*; Kiowa Valley, Scotts Bluff Co., July 28.

*The July 18 date could be incorrect, because on that day Rydberg and Flodman were on the road between Ashford and Gering, both in Scotts Bluff County.

50. ***Psoralea argophylla*** Pursh var. . . . . *sic* [*Pediomelum argophyllum* (Pursh) Grimes] – Wet meadow, Pleasant Valley, Scotts Bluff Co., July 29 [US 24014, **type**; NEB 177854, **syntype** of *Psoralea collina* Rydb.; NY 26763, July 28 [*sic*], **syntype**]. It agrees with the description of *Ps. argophylla* Pursh var. *decumbens* Gray in PACIFIC RAILROAD REPORT XII: 41, except that the stipules are very small, much shorter than the petioles. The leaves are glabrous above and slightly hairy beneath.

51. ***Psoralea campestris*** Nutt. [*Pediomelum digitatum* (Nutt. ex Torr. & A. Gray) Isely] – Sandhills, Kearney Co., June 15 [NEB 178047, 178055]; in sand, Lawrence Fork, Banner Co., July 9. [NEB 177989, 178037, 178052 are from Phelps Co. on June 18.]

52. ***Psoralea hypogaea*** Nutt. [*Pediomelum hypogaeum* (Nutt. ex Torr. & A. Gray) Rydb.] – Hills, Deuel Co. Only 3 specimens found on June 25 [NEB 177988].

53. ***Psoralea lanceolata*** Pursh [*Psoralidium lanceolatum* (Pursh) Rydb.] –Sandhills, Kearney Co., June 15; sand-draws, Lawrence Fork, Banner Co., July 9.

These dates are given in this *List* but not the *Report*.

54. ***Dalea alopecuroides*** Willd. [*D. leporina* (Aiton) Bullock] – Platte bottoms, Kearney Co., Aug. 29 [NEB 178208; NY 1270107].

55. ***Dalea laxiflora*** Pursh [*Dalea enneandra* Nutt.] – McCuligan Canyon [NY 1259327].

In July, 1890, Rydberg collected this species (also as *Rydberg 55*) on Magnesia Cliffs, Garden Co. (NEB 178078, 207900; NY 1259334).

56. ***Aurea*** T. & G. [*Dalea aurea* Nutt. ex Pursh] – Hills, Scotts Bluff Co. [August 6, NEB 178013; NY 1073405, 1073406]

July 20 and August 6 are given in the *Report* for Scotts Bluff itself, as well as August 22 for McCuligan Canyon in Deuel (today's Garden) Co. Rydberg returned to Scotts Bluff on August 6 to collect two plants that were in bud on the earlier date. He also collected this species in Garden Co. in July, 1890 (NEB 178012, 178016, 207895).

57. ***Petalostemon candidus*** Michx. [*Dalea candida* Michx. ex Willd.] –Eastern and central Nebraska; common. Platte bottoms, Kearney Co., August 29.

58. ***Petalostemon*** . . . . *sic* [*Dalea candida* Michx. ex Willd. var. *oligophylla* (Torr.) Shinners] – I collected this plant last year [1890] in Deuel [probably in what is now Garden] County [sand draws, NEB 177245, 177246]. In my notes I placed it near *P. gracile*. Dr. Britton states that it is contained in the Pringle's Collection under the name *Petalostemon candidus* var. *occidentalis* Gray, and that he cannot find any description of the same. He considers it as *P. gracile* or a variety of it, or a nearly related species. I am more inclined to refer it to *P. gracile* than to *P. candidus*, although it seems to be a connecting link. It is often prostrate, mostly commonly decumbent, sometimes ascending, rarely nearly upright (58b). The leaves are from obovate to nearly linear, obtuse, generally less than 0.5 inch long, but in one upright variety (58b) from the Platte bottoms in Scotts Bluff Co. much longer. This *Petalostemon* differs from *P. candidus* Michx. in the following points: The stem is decumbent, seldom nearly upright; the bracts are ovate with a slender point, about equaling the calyx and deciduous (in *P. candidus* longer than the calyx, and, as far as I know, persistent), leaflets obtuse (in *P. candidus* generally acute). From *P. gracile*, it differs in a stronger growth, by being more leafy, and by its broader leaves, and by the length of its spikes, which become lax in fruit, and, perhaps, by the stronger nerves on the calyx. Smaller, narrower-leaved and short spiked specimens (58c) can hardly be distinguished from *P. gracile* in the Herbarium [NEB] of the University. [Rydberg:] The full description of this *Petalostemon* is as follows:

Stem decumbent, rarely nearly upright, glabrous, striate or slightly angled; leaflets 1–2 pairs, obovate-oblong, sometimes nearly linear, 2–6 lines (rarely nearly one inch), punctate beneath; spikes cylindrical, in fruit elongated, lax; bracts deciduous, ovate, with a slender point, equaling the calyx, which is strongly nerved and glandular punctate below the lanceolate teeth; petals broadly oblong, vexillum broadly cordate.

Sandy hills; Deuel Co., June 25 [NEB 177244; RM 250565]; Lawrence Fork, July 8; Scotts Bluff, July 22 [NEB 177248; NY 1241339]; (58b) [North] Platte bottoms, Scotts Bluff Co., July 27; (58c) Kiowa Valley, July 28 [NY 1241329].

NY 1073576 is from the Platte River bottoms near Kearney City, Buffalo Co., August 29, 1891. NY 1073577 is Rydberg 58 (var. *candida*) from the South Platte River bottoms near Big Springs, Deuel Co., August 24, 1891.

In 1890, he collected this species in present-day Deuel or Garden County (NEB 177245, 177246).

J. M. Holzinger: "I made this a form of *P. candidus* [*Dalea candida* var. *oligophylla*], after close comparison with our [US] Herbarium series. With 58 I also found some *Petalostemon multiflorus*. I consider it a just source of suspicion, as to the validity of the species concerned, that *Petalostemon candidus* and *P. multiflorus* should find an almost exact parallel in *P. violaceus* and *P. tenuifolius*."

59. ***Petalostemon tenuifolius*** Gray [*Dalea purpurea* Vent. var. *purpurea* & var. *arenicola* (Wemple) Barnes] – Last year I found, near a sandy draw in Deuel Co., a few specimens of a *Petalostemon* in fruit [*Rydberg 3*, July, 1890: NEB 177192; NY 1277409, 1277411; NEB 177191 is *s.n.*, and NEB 177250 is *Rydberg 5248*]. The lowness of the plant, and the fulvous color of the spikes, caught my attention. I preserved [it] as a curious specimen of *P. violaceus*. Hunting for a description of the preceding species, I struck the description of *P. tenuifolius* in Wheeler's Report. Seeing that it fitted my other small *Petalostemon*, I gave this the name *P. tenuifolius*. There are no specimens in the Herbarium [NEB] to compare it with, but Dr. Britton, to whom I sent specimens, said that the determination was correct. This year I have collected the same in flower. The specimens fully agree with the descriptions, as well as by Dr. Watson in Wheeler's Report, as by Dr. Gray in PROC. AM. ACAD. Dr. Gray describes the flowers as being rose-purple to rose, or, in withering, to pink. The stems branch from a perennial base, are generally simple, hirsute, with long-peduncled heads. What separates it from *P. violaceus* (as I understand it) is, besides the general "habitus," the hairs of the calyx, which in *P. violaceus* are silky and more appressed and in *P. tenuifolius* more woolly and spreading, first white, then fulvous; and the shape of the bracts, which in *P. tenufolius* are broadly ovate, abruptly contracted into an awn, and in *P. violaceus* obovate or spatulate, with a broader tip and the upper part naked. Some of the specimens collected on the hills of southwestern Deuel Co. (59b) seem to be intermediate between the two, and I cannot see any specific difference between them and my specimens from Big Springs (60), which undoubtedly belong to *P. violaceus*. If any of my specimens belong to *P. tenuifolius*, I regard it as a variety of *P. violaceus*.

Sand-draw, Deuel Co., June 27 [NY 1277410], var. *purpurea*; [NY 1277675: var. *arenicola* (Wemple) Barnes]; Lawrence Fork, Banner Co., July 8 [NY 1277408, var. *purpurea*]; Kimball, August 11.

J. M. Holzinger: "*Petalostemon violaceus* Michx., with a form different from the type. In other words, I agree with you that if *P. tenuifolius* is anything at all, it is a var. of *violaceus*."

Rydberg also collected this in July, 1890, in Deuel or Garden Co. (NEB 177191, 177192, 177250).

60. ***Petalostemon violaceum*** Michx. [*Dalea purpurea* Vent. var. *purpurea*] – Common in

eastern Nebraska. Big Springs, Deuel Co., August 24.

61. ***Petalostemon macrostachyus*** Torr. var. . . . . . . *sic* [*Dalea cylindriceps* Barneby] – This species is described as having white flowers and obtuse leaves, two-thirds inch long. My specimens have yellow flowers and acute leaves, some over an inch long. *P. macrostachyus* is not given in Mr. Webber's CATALOGUE. I have seen no specimens of it. It is not found in the Herbarium [NEB] of the University, but it is reported as being in the Herbarium of Doane College, Crete, Neb. Those preserved there are said to have yellow flowers. Maybe our Nebraska plant is distinct. On the sandhills north of [North] Platte River, Scotts Bluff Co., August 4 [NEB 177086, 177088; NY 1259216, 1259217].

> J. M. Holzinger: "*Petalostemon macrostachyus* Torr. As you say, your plant has yellow flowers. But so have our specimens from Colorado. The New Mexico specimen, on the other hand, has pinkish flowers. It seems that in some few species of flowering plants the color is in unstable equilibrium. And varieties should not be made on this basis. The leaves too may vary to the extent you indicate, and do vary in our series. First descriptions, especially when based on limited material, are apt to be too rigid."

The taxonomy and precarious status of this plant was discussed by J. H. Locklear (2013: *J. Bot. Res. Inst. Texas* 7(2): 879–890), who identified the site of the holotype of *D. cylindriceps* Barneby (NY 26677) as near the South Platte River in Lincoln Co., Nebraska, in June, 1820; that specimen was collected by Edwin James of the Long Expedition.

62. ***Petalostemon villosus*** Nutt. [*Dalea villosa* (Nutt.) Spreng.] – Sand-draw, Deuel Co., August 24; sandhills, Kearney Co., August 29 [NEB 177210; NY 1277877].

In his *Report*, Rydberg spelled it both "villosus" and "villosum."

63. ***Amorpha canescens*** Nutt. – Sand-draw, Deuel Co., August 24 [NEB 176309].

64. ***Amorpha fruticosa*** L. – Near water, Platte River, Big Springs, Deuel Co., August 25.

[s.n.] ***Glycyrrhiza lepidota*** Nutt. – Too common throughout Nebraska.

65. ***Millelotus alba*** Lam. *sic* [*Melilotus albus* Medikus] – Platte bottoms, near Kearney City [Buffalo Co.], August 29 [NEB 178292, 178293].

Kearney City is in Buffalo Co., and the Platte River bottoms nearby are in both Buffalo and Kearney counties. Rydberg walked southward from the train depot in Kearney City to the Buffalo Co. side of the nearby Platte River.

66. ***Astragalus caryocarpus*** Ker. [*A. crassicarpus* Nutt.] – Prairies, throughout Nebraska. Deuel Co., June 25 [NEB 176747].

67. ***Astragalus mollissimus*** Torr. – Common westward on prairies. Roadside, Kearney Co., June 15 [NEB 176692; NY 1265815; RM 243194].

68. ***Astragalus canadensis*** L. – Throughout Nebraska. Lawrence Fork, Banner Co., July 9 [NY 1247095, 1247097]

69. ***Astragalus adsurgens*** Pall. [*A. laxmanii* Jacq.] – Sandy soil in the west. McCuligan's Canyon, Deuel [today's Garden] Co., June 26 [NY 1203643]; Lawrence Fork, Banner Co., July 8; Scotts Bluff, July 20 [NY 1203645].

Rydberg also collected this species in Banner Co. in August, 1890 (*Rydberg 49 sic*, probably 69, NY 1203644).

70. ***Astragalus racemosus*** Pursh [var. *racemosus*] – Hills: Near Curtis, Frontier Co., June 22 [NEB 176721; NY 1265591, 1265604].

71. ***Astragalus gracilis*** Nutt. – Hills and prairies in western Nebraska. Curtis, Frontier Co., June 22 [NEB 176407; NY 1249873];

Lawrence Fork, Banner Co., July 8 [NY 1249874].

Rydberg also collected this species in Deuel Co. in 1890 (NEB 207896).

72. ***Astragalus microlobus*** Gray [*A. gracilis* Nutt.] – Cultivated soil, not as common as the preceding. Deuel Co., June 25 [NEB 176421; NY 1249872 has material from Deuel Co., June 25, mounted with material from Kimball Co., August 13]; Cheyenne Co., August 13 [NY 1249875].

73. ***Astragalus missouriensis*** Nutt. [var. *missouriensis*] – There is in the Herbarium [NEB] of the University only one specimen under this name, but that belongs undoubtedly not to *A. missouriensis*, but to *A. shortianus* Nutt. Hills, Curtis, Frontier Co., June 22 [NEB 176612; NY 1262032]; [upper Lawrence Fork] North of Kimball, Kimball Co., Aug. 10 [NY 1262031].

Rydberg also collected this species in Banner Co. in August, 1890 (NEB 207887).

74. ***Astragalus pubentissimus*** Torr. & Gray (?) *sic* [*A. lotiflorus* Hook.] – I am in doubt if the determination of this species is correct. There are no specimens in the Herbarium [NEB] of the University to compare it with; the description of *A. pubentissimus* is rather brief, and besides, *Astragalus* is a very polymorphous genus. Wherefore I would not be much surprised if these specimens should belong to another species. I add here a short description:

> Plant apparently annual, low, 2-3 in., branched from the root, silky canescent; leaflets 2 pairs, oval-obovate; raceme 2-4-flowered, often very short and the flowers seemingly sessile in the axils of the leaves; flowers ochroleucous, tinged with purple; pod one-celled, membranous, ovate oblong, curved upward.

Canyon in Gosper Co.; hills near Curtis, Frontier Co., June 22 [NY 1261148, 1261149]; Lawrence Fork, Banner Co., July 9 *sic* [8 on label] [NEB 176636]; hills south of Kimball [Kimball Co.], August 12 [NEB 176636, on same sheet as the July 8 specimen].

75. ***Astragalus pictus*** Gray. var. ***filifolius*** Gray [*A. ceramicus* E. Sheldon var. *filifolius* (A. Gray) F. J. Hermann] – In fruit. Hills north of Freeport, Banner Co., July 14 [NEB 177916; NY 1247545]; Wild Cat Mountains, Banner Co., July 16.

76. ***Astragalus pectinatus*** Dougl. – In fruit; prairies near Gering, Scotts Bluff Co., July 21 [NY 1246488]; and in Pleasant Valley, Scotts Bluff Co., July 29 [NEB 176664].

77. ***Astragalus multiflorus*** [Pursh] Gray – Near a spring in Pleasant Valley [Scotts Bluff Co.] on July 29 [NEB 176969, 176704; NY 1266186; RM 223325].

78. ***Astragalus caespitosus*** Gray [*A. spatulatus* Sheld.] – In all flowering specimens I collected, the racemes were few-flowered and about of the length of the leaves. Rock, Deuel Co.; McCuligan Canyon [Garden Co.], June 26; Wild Cat Mountains, Banner Co., July 16. [NY 1283190]

Rydberg also collected this species near Lawrence Fork, Banner Co., in August, 1890 (NEB 176373; NY 1283188).

79. ***Astragalus kentrophyta*** Gray – Court House Rock, Cheyenne [now Morrill] Co., July 4 [NY 1258277; Flodman's collection on this day at this site is NEB 177011]; hills of Lawrence Fork, Banner Co., July 9 and 10; Wild Cat Mountains, Banner Co., July 16; Scotts Bluff, July 22.

NEB 177005 is from Scotts Bluff Co., August 1, 1891.

80. ***Astragalus triphyllus*** Pursh [*A. hyalinus* M. E. Jones] – Hills throughout western Nebraska. Scotts Bluff, July 22; Kiowa Valley, Scotts Bluff Co., July 27 [US 69999, **isosyntype** of *Astragalus hyalinus* Jones];

Upper Lawrence Fork, Kimball Co., August 10 [NEB 176846].

Rydberg collected this plant in Banner Co. in July, 1890 (NY 1250322) and August, 1890 (NY 5464; NEB 176847 (labelled syntype), 207917).

81. ***Astragalus sericoleucus*** Gray – Gravelly hills. In fruit only: Deuel Co., June 25 [NEB 178282; NY 1282679, 1282680].

Rydberg also collected this species in Deuel or Garden Co. in July, 1890 (NEB 178284, 207916; NY 1282681).

82. ***Oxytropis*** (?) *sic*. . . . . . [*O. multiceps* Torr. & A. Gray] – As only flowers were found I cannot decide whether it is an *Astragalus* or an *Oxytropis*. The plant is very caespitose, from a much branched, perennial caudex; leaves about 2 in., 2–3 paired, silky villous; leaflets oblong lanceolate, acute; pods (fallen among the leaves) 3–4 ", oval, membranous, one-celled, enclosed in the silky calyx; stipules oval, adnate to the petiole.–In fruit, on the hills of the Upper Lawrence Fork, Kimball Co., August 10 [NEB 177742].

J. M. Holzinger: "*Oxytropis multiceps*."

82[b]. ***Oxytropis lambertii*** Pursh – A very variable species. On the prairies, the spike is generally higher (7–10 in.) and laxer (82a) and the plant simpler. On the hills, the stem is generally more tufted, the spike lower (4–6 in.) and more dense (82b). On the barren hills near Rush Creek, Deuel [now Garden] Co., I found some specimens of a form undoubtedly belonging here. The plant (82c) is low, leaves shorter, with narrow leaflets, hirsute rather than silky; pods ½ in. long, purplish, tipped with a ½ inch long, straight, only at the end curved tip.–(82a) Kearney Co., June 15; (82b) hills south of Pumpkin Seed Creek, July 6 [NEB 176174, 176176]; (82c) hills near Rush Creek, Deuel [now Garden] Co., July 2 [NY 5224, **holotype** of *Aragallus angustatus* Rydb.]

The NEB specimens are labelled "82b", in Rydberg's hand. He also collected this in Banner Co. in August, 1890 (NY 5182, **syntype** of *Aragallus angustatus*).

83. ***Oxytropis lambertii*** Pursh. var. ***sericea*** Gray [*O. sericea* Nutt.] – In fruit on the hills near Rush Creek, Deuel [now Garden] Co., July 2, and in bloom at Scotts Bluff, July 24 [NEB 177734]. The Scotts Bluff specimens (83b) are remarkable for the size of the flowers and of the leaflets (some 1½ in. long), and for the peculiar bluish hue.

84. ***Desmodium canadense*** DC. – In fruit on the Platte bottoms, near Kearney City [Buffalo Co.], August 29 [NEB 178224, 178237].

85. ***Vic[i]a americana*** Muhl. var. ***minor*** Hook. – Common throughout Nebraska. Scotts Bluff, July 24 [NEB 207881].

86. ***Lathyrus ornatus*** Nutt. [*L. decaphyllus* Pursh] – Prairies, throughout Nebraska. Kearney Co., June 13; Gosper Co., June 20; Scotts Bluff, July 24 [NEB 177256].

87. ***Strophostyles pauciflorus*** (Benth.) Wats. [*S. leiosperma* (Torr. & A. Gray) Piper] – In Gray's Manual, the leaflets and the pods are said to be 1 inch long. In the specimens gathered on the sand of the [North] Platte River, near Horse Creek, Scotts Bluff Co. [August 1, NEB 177693], both are about 1½ in. long. In Coulter's Manual it is not given at all.

88. ***Desmanthus brachylobus*** Benth. [*D. illinoensis* (Michx.) MacM. ex B. L. Robins.] – Platte bottoms, near Kearney City [Buffalo Co.], August 29 [NEB 176033, 176034].

89. ***Schrankia uncinata*** Willd. [*Mimosa quadrivalvis* var. *nuttallii* (DC.) Beard ex Barneby] – Hillsides and canyons in Gosper Co., near Bertrand, June 20 [NEB 176068, 176078].

Hapeman took Rydberg from Bertrand, in Phelps Co., into adjacent Gosper Co.

ROSACEAE

90. ***Prunus demissa*** Walp. [*P. virginiana* L., *s. l.*] – Hills and canyons: Gosper Co., June 20;

Curtis, Frontier Co., June 22; Lawrence Fork, Banner Co., July 3.

> In his *Report*, Rydberg cited this plant at various sites on June 20, 22, 26; July 3, 8, 9; and August 14, 17, and 19.

91. ***Prunus pumila*** L. [var. *besseyi* (L. H. Bailey) Waugh; *P. besseyi* L. H. Bailey] – Sand-draws of western Nebraska. It is not given in Coulter's MANUAL. Lawrence Fork, Banner Co., July 8 [NEB 175883].

92. ***Cercocarpus parvifolius*** Nutt. [*C. montanus* Raf.] – Collected for the first time in Nebraska by me in 1890 [Banner Co., NEB 175023, 175024, 175027]. On the hills between Pumpkin Seed Valley and Lawrence Fork, Banner Co., July 9 *sic* [10 on label of NEB 175025], and on the Wild Cat Mountains, July 16 [RM 224075]. It was also seen in Cedar Canyon [Scotts Bluff Co.] and near Scotts Bluff.

93. ***Geum album*** Gmel. [*G. canadense* Jacq.] – Not given in Coulter's MANUAL. Among bushes in Lawrence Fork, Banner Co., July 9.

94. ***Geum strictum*** Ait. [*G. aleppicum* Jacq.] – Among bushes in Lawrence Fork, Banner Co., July 9 [NEB 175205]; north of Scotts Bluff, July 21 [NEB 175202].

> Spelled "strictum" on the NEB specimen labels and in most places in the *Report*, but "striatum" for July 9 in the *Report*.

95. ***Potentilla supina*** L. [*P. paradoxa* Nutt.] – On Platte bottoms north of Scotts Bluff, July 21 [NEB 175330]; and near Horse Creek, Scotts Bluff Co., August 1 [NEB 175336].

96. ***Potentilla pensylvanica*** L. – Prairies: Deuel Co., June 25.

97. ***Potentilla anserina*** L. – Platte bottoms, Scotts Bluff, July 21, and Horse Creek, August 1 [NEB 175019].

98. ***Rosa blanda*** Ait. [*R. woodsii* Lindl.] – I am not satisfied with the determination of the roses. The descriptions seem to be in a confusion. Lobed sepals and not lobed sepals, narrow stipules and dilated stipules are made essential specific characters. In fact, the contrary characters are often found in the same individual. *R. blanda* and *R. Arkansana* seem to be only varieties of the same species; so also *R. Fendleri* and *R. Woodsii*. *R. blanda* is found in Nebraska only in the west. Canyon, south of Pumpkin Seed Valley, Banner Co., July 6 [NEB 175426].

> *Rosa woodsii* is common in central and western Nebraska, but the presence of *R. blanda* anywhere in Nebraska, as interpreted in FLORA OF NEBRASKA (Kaul et al., 2011) and FLORA OF NORTH AMERICA 9 (Lewis et al., 2014), is doubtful.

99. ***Rosa arkansana*** Porter – Prairies: Curtis, Frontier Co., June 22 [NEB 175520]. Some stipules on the specimens are rather broad.

100. ***Rosa nutkana*** Presl [*R. woodsii* Lindl.] – Not reported for Nebraska before. Low ground, Curtis, Frontier Co., June 22 [NEB 175453]; McCuligan Canyon, Deuel [now Garden] Co., June 26. In my specimens, some stipules are narrow and the flowers are smaller than described in Coulter's MANUAL.

> *Rosa nutkana*, as now interpreted (Lewis et al., 2014, in FLORA OF NORTH AMERICA 9), does not occur east of the Rocky Mountains.

101. ***Rosa fendleri*** Crepin [*R. woodsii* Lindl.] – Curtis, Frontier C., June 22; sand-draw north of Potter, Cheyenne Co., August 17 [NEB 175461].

102. ***Rosa woodsii*** Lindl. – Although I do not regard this a separate species, I do label it *R. Woodsii*, as in most cases some of the sepals have 1-few linear lobes, a quality that is, by far, not constant. Canyons of the west: Lawrence Fork, Banner Co., July 8 [RM 224161B]; Wild Cat Mountains, July 16 [RM 224161A]; Scotts Bluff, July 29 *sic* [July 22 on label of NEB 175476].

103. ***Amalanchier** sic **alnifolia*** Nutt. – Hillsides: Lawrence Fork, Banner Co. and in a canyon west of Big Horn Mt., Banner Co., August 8 (only leaves) [NEB 175703].

104. ***Ribes cereum*** Dougl. [var. *inebrians* (Lindl.) C. L Hitch.] – Hills: Lawrence Fork, Banner Co., July 8 [NEB Co., July 16.

In July, 1890, Rydberg collected this species from Deuel Co. (NEB 79564) and from McCuligan Canyon in Deuel (today's Garden) Co. (NEB 207908), and in August from Banner Co. (NEB 174949).

105. ***Ribes floridum*** L. [*R. americanum* P. Mill.] – Spring in Kiowa Valley, Scotts Bluff Co., July 28.

106. ***Ribes aureum*** Pursh [*R. odoratum* H. L. Wendl.] – The fruit is very variable: 106a–Black, spherical–the most common form, July10. 106b–Yellow, spherical. Canyon near Scotts Bluff, July 21 [US 691782**, type** (or syntype; US 590343, a duplicate) of *Ribes aureum* var. *chrysococcum* Rydb.] 106c–Black, ellipsoid, large. [North] Platte bottoms, Scotts Bluff Co., July 21 [NEB 79514, as 106b].

Rydberg collected the black-fruited form in Banner Co., August, 1890 (NEB 174863, 207909).

ONAGRACEAE

107. ***Epilobium adenocaulon*** Haussknecht [*E. ciliatum* Raf.]–Very variable. Lawrence Fork, Banner Co., July 8; Carter Canyon, July 25; Spring, Kiowa Valley, July 28; Lodgepole Creek, near Kimball, August 12 [NEB 180920]; Sidney, August 19; 107b–A simpler form with narrower leaves, from Horse Creek [Scotts Bluff Co.], Aug. 1 [NEB 180916].

108. ***Oenothera biennis*** L. var. . . . . . *sic* – Flowers unusually small. Scotts Bluff, July 24; Kiowa Valley, Scotts Bluff Co., July 28 [NEB 181216].

In July, 1890, Rydberg collected this species in Deuel Co. (NEB 181176).

109. ***Oenothera sinuata*** L. [*O. laciniata* Hill] – This is not given in Coulter's MANUAL. A few small specimens were found in a sand-draw in Deuel Co. June 27 [NEB 181270, US 56675].

110. ***Oenothera pinnatafida** sic* Nutt. [*O. albicaulis* Pursh, *O. albicaulis* Pursh var. *nuttallii* (Sweet) Engelm.] – Very variable, from simple to much branched below; leaves simple or pinnatifid. Curtis, Frontier Co., June 22 [NEB 180851, 180852]; sandy prairies in Deuel Co., June 27; Rush Creek [Garden Co.], July 2; Kimball, August 11.

NY's online list shows "110. *Carduus nebraskensis* Britton, Scotts Bluff, on July 24, 1891, NY 163071, holotype", but the labels on NEB's specimens and in Rydberg's original *List* manuscript show 110 as *Oenothera pinnatifida*, and *Carduus nebraskensis* (now *Cirsium canescens* Nutt.) is not in the *List*. However, he did collect in that area on that day.

111. ***Oenothera albicaulis*** Nutt. [*O. nuttallii* Sweet, *O. albicaulis* Pursh var. *nuttallii* (Sweet) Engelm.] – Prairies; Deuel Co., June 25 [NEB 180820, 180829]; Pumpkin Seed Valley, Banner Co., July 16 [US 44483]; Cedar Canyon, Scotts Bluff Co., July 20; Kimball, August 12.

112. ***Oenothera albicaulis*** Nutt. var. . . . . .*sic* [*O. pallida* ssp. *latifolia* (Rydb.) Munz, *fide* W. L. Wagner, 2012] – Low, diffusely branched, silky puberulent; leaves oblong lanceolate, toothed; calyx-teeth sometimes not free. It seems to be very near to var. *sinuata* Wats. It grows, wherever I have seen it, in the sand-draws, when the type with the linear, glabrate leaves always is found on the prairies. Deuel Co., June 25 [US 44482]; Upper Lawrence Fork, Kimball Co., Aug. 10.

J. M. Holzinger: "I make this simply *Oenothera albicaulis*."

Rydberg also collected this species in Deuel Co., July, 1890 (NEB 207907).

113. ***Oenothera coronopifolia*** T. & G. – Sandy soil: Deuel Co., June 25 [NEB 180770];

Lawrence Fork, Banner Co., July 10 [US 44484]

J. M. Holzinger: "*Oenothera caespitosa.*"

114. ***Oenothera caespitosa*** Nutt. var...... *sic* – The capsule is not oblong, but rather narrowly ovate, the hairiness shorter than that of *Oe. marginata* (=*Oe. caespitosa* Nutt.) in the Herbarium [NEB] of the University. Canyons of Scotts Bluff, July 22 [NEB 181385, US 55901].

J. M. Holzinger: "I make this simply *Oenothera caespitosa.*"

115. ***Oenothera canescens*** Torr. – Dr. Watson in his Revision describes *Oe. canescens* as having linear or narrow lanceolate leaves. My specimens have from lanceolate to ovate leaves. Hills: Deuel Co., June 27 [NEB 181117, US 56677].

In 1890, Rydberg collected it in Banner Co. (NEB 181116).

116. ***Oenothera serrulata*** Nutt. – Prairies: Deuel Co., June 25 [NEB 181331]; Lawrence Fork, Banner Co., July 10.

117. ***Oenothera hartwegi*** Benth. var. ***lavandulaefolia*** Wats. [*O. lavandulifolia* Torr. & A. Gray; *Calylophus lavandulifolius* (Torr. & A. Gray) P. Raven] – Hills: Lisbon, Perkins Co., June 23; Deuel Co., June 25 [US 44060] and 27; Rush Creek [Garden Co.], July 2; Pleasant Valley, Scotts Bluff Co., July 28 [NEB 180982].

Rydberg also collected it in August, 1890, in Banner Co. (NEB 207892)

118. ***Gaura coccinea*** Nutt. – Prairies: Kearney Co., June 15 [NEB 181010, 181018, 181039]; Pumpkin Seed Valley, Banner Co., July 14.

119. ***Gaura coccinea*** Nutt. var...... *sic* – Glabrous throughout; stem light brown and sheddy; flowers light pink. Prairies: Deuel Co., June 27 [NEB 181031].

120. ***Gaura parviflora*** Dougl. – [North] Platte bottoms in Cheyenne [now Morrill] Co., July 3.

Modern Cheyenne County is much reduced from its 1891 size and is not on either the North or South Platte rivers, but the North Platte River flows through today's Morrill Co., which was carved out of the Cheyenne Co. of Rydberg's time.

In July, 1890, Rydberg collected this species near Lodgepole Creek and Chappell, both in Deuel Co. (NEB 181078, 207906).

LOASACEAE

121. ***Mentzelia*** sp...... *sic* [*M. albicaulis* (Dougl. ex Hook.) Dougl. ex Torr. & A. Gray] – I do not know whether to refer to *M. albicaulis* or to *M. dispera* Wats. The leaves are lanceolate or even ovate, entire or rarely sinuately toothed; the seeds are obtusely angled and finely muricate.–North side of Scotts Bluff, about 5,300 feet above the sea, July 23 [NEB 180076, 180437, 180438, 180439].

J. M. Holzinger: " Should be entered as *Mentzelia albicaulis.*"

122. ***Mentzelia ornata*** T. & G. [*M. decapetala* (Pursh) Urban] – Canyons, Scotts Bluff, July 24 [NEB 180443].

123. ***Mentzelia nuda*** T. & G. – This plant has been collected for several years, and from different localities in Nebraska. It has been named *M. ornata* and *M. laevicaulis*; on one hand, because the calyx is bracteolate, on the other hand because it is often found open in the afternoon. All Nebraska specimens I have seen, as well in collections as in the field, have the calyx bracteolate; and I would have regarded them as belonging to a distinct species, had not Mr. Porter in the FLORA OF COLORADO made the statement that *M. nuda* has bracteolate calyx. The only specimens that come near to the description of *M. nuda* in this respect are those stunted ones collected by me on the hills of Upper Lawrence Fork, Kimball Co., August 10 (123b); but even in these are the bracts present, although one one or two,

small and linear. In these are the pods also unusually short. Canyons and sand-draws: Lawrence Fork, Banner Co., July 10; Wild Cat Mountains, Banner Co., July 16 [NEB 180471]; Kiowa Valley, Scotts Bluff Co., July 28 [NEB 180407]; McCuligan Canyon, Deuel [now Garden] Co., Augst 22.

J. M. Holzinger: "*Mentzelia nuda!*"

Rydberg also collected this species in Deuel or Garden Co., July, 1890 (NEB 180476, 180479, 207905).

CUCURBITACEAE

219. *sic* [123 on specimen label]. ***Cucurbita foetidissima*** H.B.K. – Lawrence Fork, Banner Co., July 9 [NEB 186919].

219 is also *Stephanomeria minor* in Rydberg's *List*.

CACTACEAE

[*s.n.*] ***Mammil[l]ari[a] vivipara*** Haw. [*Coryphantha vivipara* (Nutt.) Britt. & Rose] – It was collected on the sandhills of Kearney Co., but only one specimen [NEB 180531, Rydberg *s.n.*] was preserved and is deposited in the Herbarium [NEB] of the University.

124. [126* on specimen label] ***Opuntia missouriensis*** DC. [*O. polyacantha* Haw. var. *polyacantha*] – Platte bottoms, Phelps Co., June 18 [NEB 180524, US 3046542]

*Erroneously posted online at US as Rydberg 926, an apparent misreading of 126. Rydberg's 1's and 9's are sometimes difficult to distinguish.

125. ***Opuntia rutila*** Nutt. [*O. polyacantha* Haw. var. *polyacantha*] – New to Nebraska. – It was collected on the prairies of Deuel Co., June 27 [25 on label] [NEB 180521, 180526; US 3046540]. A yellow-flowered variety (125b) [NEB 180494, 180527] was growing together with the purple one.

J. M. Holzinger: "*Opuntia rutila* Nutt. I do not know this plant, and from the material you sent, as well from the meager descriptions, I am left in doubt whether this is really *O. rutila* of Nuttall. I am inclined to think that it is a variety of *Opuntia missouriensis*."

126. ***Opuntia fragilis*** Haw. – Prairies, Deuel Co., June 27. [NEB 180497, US 3046541 as *O. fragilis*, 3046821 as *Opuntia* sp.]

J. M. Holzinger: "It is not *Opuntia fragilis*, but the material is too scant for full judgment."

UMBELLIFERAE

127. ***Musenium*** *sic* ***tenuifolium*** Nutt. [*Musineon tenuifolium* Nutt. ex Torr. & A. Gray] – There is no doubt in my mind that this is *M. tenuifolium* Nutt., although the flowers are yellow. In withering they become light straw-color, which undoubtedly made Mr. Nuttall believe that they had been white. The fruit is of the *Musenium* type. It must be referred to this genus, even if it is not *M. tenuifolium* Nutt. It is common on the hills of Banner and Scotts Bluff counties, and on Court House Rock, Cheyenne [now Morrill] Co. It gave a good illustration of what influence the altitude has on the time of flowering. On July 4, we collected on Court House Rock, Cheyenne [now Morrill] Co., specimens in well-developed fruit, but only one specimen was found in flower. Nearly a month later, on July 28, we found it in full bloom on the hills of Pleasant Valley, between 40 and 50 miles further west, and only 300 or 400 feet higher above the sea [NEB 181734]. [NEB 181628: Pumpkin Cr., Scotts Bluff Co., July 6, 1891.]

In August, 1890, Rydberg collected this species in Banner Co. (NEB 181621, 181625).

128. ***Berula angustifolia*** Koch [*B. erecta* (Huds.) Cov. var. *incisa* (Torr.) Cronq.] – Springs: Lawrence Fork, Banner Co., July 10; Kiowa Valley, Scotts Bluff Co., July 28 [NEB 181486].

Rydberg also collected this species in Banner Co. in August, 1890 (NEB 181480, 181490).

129. ***Pseudocymopterus anisatus*** Coulter & Rose [*Lomatium nuttallii* (A. Gray) J. F. MacBr.] – Collected for the first time in Nebraska. The rays are sometimes more than twelve, and the fruit over 3 lines long. Scotts Bluff Co., in the Pass, July 22 [NEB 181619; NY 1141437, as *L. gr*aveolens (S. Watson) Dorn & L. Hartman].

J. M. Holzinger: *"Peucedanum kingii."*

130. ***Cymopterus montanus*** T. & G. – Dooryard, Deuel Co., June 27.

131. ***Cicuta maculata*** L. – Platte bottoms, Kearney Co., June 15; Scotts Bluff Co., July 20 [NEB 181567].

CORNACEAE

132. ***Cornus stolonifera*** Michx. [*C. sericea* L.] – Low land: Platte bottoms, Kearney Co, June 15; canyon, Kiowa Valley, Scotts Bluff Co., July 28.

CAPRIFOLIACEAE

133. ***Symphoricarpos occidentalis*** Hook. – The style is often in this species slightly bearded, a fact that Rev. Bates of Valentine has called my attention to. Lawrence Fork, Banner Co., July 8, and near Sidney [modern Cheyenne Co.], August 19.

RUBIACEAE

134. ***Galium boreale*** L. – Canyon north of Pumpkin Seed Valley, Banner Co., July 11 [NEB 186541, 207811, isotypes of *G. b.* var. *linearifolium* Rydb.; NY 131422 **holotype** of *G. b.* var. *linearifolium* Rydb.**]**; Wild Cat Mountains, Banner Co., July 16 [NEB 181531, 186543]; spring, Pleasant Valley, Scotts Bluff Co., July 29 [NEB 207812]

Rydberg also collected this species in Banner Co. in August, 1890 (*Rydberg 81*, NEB 186534).

COMPOSITAE

135. ***Vernonia fasciculata*** Michx. – Platte bottoms, Big Springs, Deuel [!] Co., August 24 [NEB 191663].

On July 14, 1900, Rydberg collected this species in Kearney Co., near the Platte River (NY 274482, **holotype** of *V. fasciculata* var. *nebraskensis* Gleason).

136. ***Kuhnia eupatorioides*** L. var. ***corymbulosa*** T. & G. [*Brickellia eupatorioides* (L.) Shinners var. *corymbulosa* (Torr. & A. Gray) Shinners] – Carter Canyon, Scotts Bluff Co., July 25 [NEB 190011].

Rydberg also collected it in Banner Co. in August, 1890 (NEB 190003, 190020, 190030, 207889).

137. ***Brickelia*** *sic* ***grandiflora*** Nutt. – Collected for the first time in Nebraska, by me in 1890 [*s.n.*, Banner Co., NEB 188780, 188785, 207629]. It was seen in several new localities this summer. Canyon south of Lawrence Fork, Banner Co., July 10 [NEB 188778, 207813; RM 221986]; Scotts Bluff, July 22.

138. ***Liatris punctata*** Hook.–Throughout Nebraska. Scotts Bluff , August 3; McCuligan Canyon, Deuel [now Garden] Co., August 22 [NEB 190248].

139. ***Liatris spicata*** Willd. [*L. lancifolia* (Greene) Kittell] – My specimens are like *L. pycnostachya* Michx. in that the bracts are colored, but they are obtuse and appressed as in *L. spicata*. The flowers are about 10 in each head.–[North] Platte bottoms, west of Scotts Bluff, August 3 [NEB 190341]. It is not reported for Nebraska before, and it is not given in Coulter's MANUAL.

140. ***Gutierrezia euthamiae*** T. & G. [*G. sarothrae* (Pursh) Britt. & Rusby] – Very variable; nearly every locality has its own form. Canyons, near Scotts Bluff, July 27 [NEB 189742, 207814]; prairie near Horse Creek [Scotts Bluff Co.], August 1 [NEB 189729].

In 1890, Rydberg collected this species in Banner and Deuel counties (NEB 189736, 207929).

141. ***Grindelia squarrosa*** Dunal – The specimens seen in western Nebraska had unusually thin leaves. Carter Canyon, Scotts Bluff Co., July 25; Horse Creek, August 1 [NEB 189688].

142. ***Applopappus armerioides*** Gray [*Stenotus armerioides* Nutt.] – I am in doubt if the determination of this is correct, owing to the lack of good material and to the fact that there are no specimens of *A. armerioides* or of any related species in the Herbarium [NEB] of the University. *A. armerioides* is not recorded for Nebraska. Hill, south of Freeport, Banner Co., July 6 [NEB 191406; NY 260046 is Rydberg *s.n.*]

This plant is now known from six counties in western Nebraska.

143. ***Applopappus nuttallii*** T. & G. [*Xanthisma grindelioides* (Nutt.) Morgan & Hartm.] – Not in Webber's CATALOGUE. I collected only one species, in 1890, but this year I found that it is very common in Banner, Scotts Bluff, and Kimball counties. Hills south of Freeport, Banner Co., July 6 [NEB 190781, 190783, 207815]; Wild Cat Mountains, July 18; Upper Lawrence Fork, Kimball Co., August 11 [NEB 207820].

Rydberg also collected this species in Banner Co. in August, 1890 (NEB 190779).

144. ***Applopappus spinulosus*** DC. [*Machaeranthera pinnatifida* (Hook.) Shinners, *Xanthisma spinulosum* (Pursh) Morgan & Hartm.] – Prairies, common: Kearney Co., June 18; Wild Cat Mountains, Banner Co., July 16 [NEB 190820, 190822]; Upper Lawrence Fork, Kimball Co., August 11.

NEB 190809 is from the Rush Creek area, Garden Co., July 2, 1891.

145. ***Applopappus rubiginosus*** T. & G. [*Rayjacksonia annua* (Rydb.) Hartm. & Lane, *Haplopappus annuus* (Rydb.) Cory] – Platte Valley, Julesburg [Sedgwick Co.], Colorado, August 28 [NEB 128215, 128216, 128218, 207816].

146. ***C[h]rysopsis villosa*** Nutt. [*Heterotheca villosa* (Pursh) Shinners] – Common.–Pumpkin Seed Valley, July 14 [NEB 207821]; Kiowa Valley, Scotts Bluff Co., July 28 [NEB 188875].

In 1890, Rydberg collected this species in July in Deuel Co. (NEB 188805, 188846, 188847), and in August in Banner Co. (NEB 188850).

147. ***Crysopsis*** [*sic, Chrysopsis*] ***villosa*** Nutt. var. ***hispida*** Gray [*Heterotheca villosa* (Pursh) Shinners] – Sand: Lawrence Fork, Banner Co., July 10 [NEB 131631]. Some of the specimens from the hills south of Pumpkin Seed Valley, July 11 [NEB 188804], seem to be between this and var. *ericoides*.

Rydberg collected this species in Deuel Co. in July, 1890 (NEB 188805).

149 [*sic*, 148 no doubt intended] ***Crysopsis villosa*** Nutt. var. ***canescens*** Gray [*Heterotheca villosa* (Pursh) Shinners] – McCuligan Canyon, Deuel [now Garden] Co., August 22 [NEB 188874]; Lawrence Fork, Banner Co., July 8; hills north of [North] Platte River, Scotts Bluff Co. August 5.

149. ***Crysopsis villosa*** Nutt. var. ***sessiliflora*** Gray [*Heterotheca villosa* (Pursh) Shinners] – Collected for the first time in Nebraska. Sand-draw, Deuel Co., August 24 [NEB 188813].

150. ***Bigelovia howardi*** *sic* Gray [E*ricameria parryi* (A. Gray) Nesom & Baird var. *howardii* (Parry ex A. Gray) Nesom & Baird] – Not reported for Nebraska. Hills, north of Freeport, Banner Co., July 14 (not in bloom); south of Scotts Bluff, July 24 [NEB 188888; NY 214162 (not a type) is mounted on same sheet as Linosyris howardii NY 214161 above].

151. ***Bigelovia graveolens*** var. ***glabrata*** Gray [*Ericameria nauseosus* (Pallas ex Pursh) Nesom & Baird var. *graveolens* (Nutt.) Reveal

& Schuyler] – 5-7 feet high bush. Common, east of Scotts Bluff, July 21 [NEB 188869] and August 6.

152. ***Solidago missouriensis*** Nutt. – It is very variable; several forms were collected: **a**:– Kiowa Valley, Scotts Bluff Co., July 28 [NEB 191035]; McCuligan Canyon, Deuel [now Garden] Co., August 22 [NEB 191006]; **b**: Cedar Valley, Scotts Bluff Co., July 25 [NEB 190966, 191004] ; **c**: Kimball, August 12 [NEB 190975].

J. M. Holzinger: "All [152, 153, 154] typical. All *Solidago missouriensis*, in my judgment. A variable species, passing eastward into *S. juncea*, to which your narrow-leaved form is an approach. There has come through our hands only one very distinct variety of this species, from Indian Territory, with fascicles of leaves in the leaf axils, which has been named *S. missouriensis* var. *fasciculata*."

Rydberg also collected the species in Banner Co., August, 1890 (NEB 190979, 190987, 190991, 190999).

153. ***Solidago missouriensis*** Nutt. . . . . . *sic* – I have given this a separate number because it differs considerably from the type. The panicle is unusually narrow; the bracts few, long, thin, and papery margined. Near Scotts Bluff, July 25 [NEB 190969, 207831].

154. ***Solidago missouriensis*** Nutt. var. . . . . . *sic* – Leaves thin, lanceolate to nearly linear; stem very slender, panicle small, narrow, achenes somewhat silky. Canyon, Kiowa Valley, Scotts Bluff Co., July 27 [NEB 207829].

155. ***Solidago canadensis*** L. – Common; [North] Platte River, west of Scotts Bluff, August 3 [NEB 207830].

156. ***Solidago serotina*** Ait. (?) *sic* [*S. gigantea* Aiton] – This is perhaps only a smooth variety of *S. canadensis*. [North] Platte River, north of Scotts Bluff, July 25 [NEB 191211].

157. ***Solidago nemoralis*** Ait. var. ***incana*** Gray [*S. mollis* Bart.] – Common in the west. Hills: near Kimball, August 12; in Kiowa Valley, Scotts Bluff Co., July 28 [NEB 191068].

J. M. Holzinger: "Is *Solidago radula* Nutt."

158. ***Solidago nemoralis*** Ait. var. . . . . . *sic* [*S. mollis* Bart.] – This seems to be very near to *S. nemoralis*, but the leaves are elliptic oblong to obovate, entire or slightly toothed, very scabrous; panicle narrow, leafy, the branches short and nearly upright. It seems to connect *S. nemoralis* and "GLOMERULARFLORAE." Hills near Lodgepole, Cheyenne Co., August 19 [NEB 191069, 191073].

J. M. Holzinger: "Is the same [*S. radula*] with stunted panicle."

Rydberg collected the species in Deuel Co., July, 1890 (NEB 191065, 191079, 191087).

159. ***Solidago lanceolata*** L. [*Euthamia gymnospermoides* Greene]–Throughout Nebraska. – [South] Platte River, Big Springs, Deuel County, August 24 [NEB 189560].

Solidago rigida was seen at several places, but not in bloom

160. ***Townsendia grandiflora*** Nutt. – Hills of western Nebraska: North of Pumpkin Seed Valley, Banner Co., July 6 [NEB 207832].

161. ***Aster commutatus*** T. & G. [*A. falcatus* subsp. *commutatus* (Torr. & A. Gray) A. G. Jones] – The heads of some of the specimens are rather small, but contain over 20 rays. Near water: Carter Canyon, Scotts Bluff Co., July 25; Lodgepole Creek, near Kimball, August 12 [NEB 207897]; sand-draw in Deuel Co., August 24.

162. ***Aster adscendens*** *sic* Lindl. (?) *sic* – I am in doubt if the determination of this and of the following is correct. Lodgepole Creek, near Kimball [Kimball Co.], August 12. [Identity and location of specimen unknown.]

The only Nebraska locality now known for *A. ascendens* is Sioux Co., more than 100 miles away.

163. ***Aster fremontii*** Gray – Horse Creek, Scotts Bluff Co., August 1.

The name is synonymous with *A. spathulatus* Lindl., a montane species unknown in Nebraska. There is no specimen at NEB, and the issue remains unresolved..

164. ***Aster tradescantii*** L. – Platte bottoms, Big Springs, Deuel [!] Co., August 24.

According to FNA, *Aster tradescantii* may be confused with *A. ontarionis* Wieg., but that plant does not occur closer than 300 miles to the east. *Aster tradescantii* is a plant of rocky shores and streamsides in extreme eastern states and provinces. The location and identity of this specimen remain unresolved.

165. ***Aster foliaceus*** Lindl. – [North] Platte [River], north of Scotts Bluff, July 23; Horse Creek, Scotts Bluff Co., August 1; Lodgepole Creek, near Kimball, August 12 [NY 161978, **holotype** of *Aster phyllodes* Rydb.; no specimen at NEB]

NY 161978 was annotated as *A. foliaceus* by A. Cronquist in 1940, M. L. Dean in 1964, and A. G. Jones in 1980, all annotators indicating its hybrid origin with species unknown in Nebraska. There are no specimens at NEB, and the species has not been found again in Nebraska, where there is no appropriate montane habitat. This issue remains unresolved.

166. ***Aster ericaefolius*** Rothrock [*Chaetopappa ericoides* (Torr.) G. L. Nesom] – Prairies: Venango, Perkins Co., June 23.

167. ***Aster tanacetifolius*** H.B.K. [*Machaeranthera tanacetifolia* (Kunth) Nees] – Common in the west. Scotts Bluff, July 23.

168. ***Erigeron pumilus*** Nutt. – My specimens are unusually large. Hills: Deuel Co, June 27 [NEB 189412, 207833]; Lawrence Fork, Banner Co., July 10.

169. ***Erigeron canus*** Gray – Not uncommon, McCuligan Canyon, Deuel [now Garden] Co., June 25 [NEB 189345]; Scotts Bluff, July 22 [NEB 189356].

170. ***Erigeron divergens*** T. & G. [*E. bellidiastrum* Nutt.] – Scotts Bluff, July 22 [NEB 189427, 207809].

Rydberg collected this in Banner Co. in July, 1890 (NEB 207927) and August, 1890 (NEB 189358, 189359, 189424). The label for NEB 189359 has "5876" written on it, but that is Herbert Webber's numbering system for his private herbarium, not Rydberg's field number.

171. ***Erigeron strigosus*** Muhl. – Sandhills, Kearney Co., June 15 [NEB 189448, 189464, 189466].

172. ***Erigeron canadensis*** L. [*Conyza canadensis* (L.) Cronq.] Not very common in the west. Canyon, Kiowa Valley, Scotts Bluff Co., July 28.

173. ***Antennaria plantaginifolia*** Hook. [*A. microphylla* Rydb.] – An unusually slender and small-leaved form from Pumpkin Seed Valley, Banner Co., July 16 [NEB 188257].

174. ***Iva ciliata*** Willd. [*I. annua* L.] – Platte [River], Kearney City [Buffalo Co.], August 29* [NEB 190516, 207807].

* The labels on the specimens note August 27.

175. ***Iva axillaris*** Pursh – Not in Mr. Webber's Catalogue. Near [North] Platte River, Scotts Bluff Co., July 23 [NEB 190499, 190500].

176. ***Iva xanthifolia*** Nutt. [*Cyclachaena xanthiifolia* Fresen.] – Scotts Bluff, July 24. [NEB 190533 is from this site on Aug. 5, 1891.]

177. ***Ambrosia trifida*** L. var. ***integrifolia*** T. & G. – Rare in eastern Nebraska, but more common than the type in the west. Sand-draw near Potter, Cheyenne Co., August 15 [NEB 188030].

Rydberg also collected this species at Hackberry Springs, Banner Co., in August, 1890 (NEB 188032).

178. ***Ambrosia psilostachya*** DC. – Rare in eastern Nebraska, but more common in the western part of the state. Sand-draw near Potter, August 15 [NEB 187996].

179. ***Franseria hookeriana*** Nutt. [*Ambrosia acanthicarpa* Hook.] – It is not given in Mr. Webber's CATALOGUE OF THE FLORA OF NEBRASKA. Seemingly, it was collected for the first time in Nebraska this year. Maybe it has been overlooked on account of its likeness to *Ambrosia psilostachya* and *A. artemisaefolia*. Sand-draw near Potter, Cheyenne Co., August 15 [NEB 189654, 189657, 189659, 189660; RM 2648].

180. ***Franseria discolor*** Nutt. [*Ambrosia tomentosa* Nutt.] – There is no record at the University [NEB] that it has been collected in Nebraska, except that it is given for the state in the SYNOPTICAL FLORA and consequently also in Coulter's MANUAL. Valley 6 miles south of Dix, Kimball Co., Aug. 14 [NEB 189643, 189645, 207898; RM 2647].

181. ***Franseria tomentosa*** Gray [*Ambrosia grayi* A. Nels.) Shinners] – It has not been reported for Nebraska before, although it already begins to become a troublesome weed in Kearney Co., especially on low, rich soil. Near Axtell, August 29 [NEB 133898, 189664].

182. ***Lepachys columnaris*** T. & G. [*Ratibida columnifera* (Nutt.) Woot. & Standl.] – Throughout Nebraska. Cedar Valley, Scotts Bluff Co., July 16 [NEB 190437].

Rydberg collected this species in August, 1890, in Banner Co. (NEB 207890).

183. ***Lepachys columnaris*** T. & G. var. ***pulcherrima*** T. & G. [*Ratibida columnifera* (Nutt.) Woot. & Standl.] – Varies, with brown rays, Cedar Valley, Scotts Bluff Co., July 18 [NEB 190404]; and with purplish rays, Scotts Bluff , July 22 [NEB 190402].

184. ***Helianthus annus*** *sic* L. var. . . . . .*sic* – Small specimens in appearance more like H. petiolaris. Near Kimball, August 12.

185. ***Helianthus petiolaris*** Nutt. – Prairies, Kimball Co., August 12 [NEB 189927].

186. ***Helianthus rigidus*** Desf. [*H. pauciflorus* Nutt. ssp. *subrhomboideus* (Rydb.) O. Spring & E. Schilling] – Throughout Nebraska on the hills. Upper Lawrence Fork, Kimball Co., August 10 [NEB 189960].

187. ***Helianthus maximiliani*** Schrader. – Platte bottoms, Kearney City [Buffalo Co.], August 29 [NEB 199111].

188. ***Helianthus maximiliani*** Schrader var. . . . . . *sic* – This is undoubtedly *H. maximiliani*, although the lower part of the stem is smooth, the rays are golden yellow, and the paleae are long, lanceolate. It may be a broad-leaved and hairy variety of *H. nuttallii*, which I have never seen. [North] Platte River near Horse Creek, Scotts Bluff Co., August 1.

J. M. Holzinger: "I could make this only a form of *H. maximiliani*."

189. ***Coreopsis tinctoria*** Nutt. – Upper Lawrence Fork, Kimball Co., August 10 [NEB 189121].

190. ***Bidens cernua*** L. var. . . . . . *sic* – A low discoid form, with nearly entire leaves, large outer involucre and hardly nodding head. Sands of South Platte River, Julesburg [Sedgwick Co.], Colorado, August 28 [NEB 130283, 130294, 207806; NEB 188738 is from Lawrence Fork, Banner Co., August 10].

J. M. Holzinger: "190, 191. In these plants too, allowance must be made for variation; I place your plants in the respective species."

In 1890, Rydberg collected this species in Banner Co. (NEB 188732).

191. ***Bidens chrysanthemoides*** Mich. var. . . . . .[*B. cernua* L.] – This is the *B. chrysanthemoides* of several Nebraska collectors and of L. F. Ward from Colorado. In many respects, it does not agree with the description of *B. chrysanthemoides* in Gray's

and Coulter's manuals. The outer involucre is longer than the inner, sometimes very foliaceous; the rays are only about ½ longer than the inner bracts, oval, about ½ in., long, golden yellow; leaves rather sharply serrate, a little connate; heads little nodding, if at all. *Bidens cernua* is common in eastern Nebraska, especially the discoid form. When well developed, it always has nodding head, with large bracts. The only (in my opinion) true *B. chrysanthemoides* I have seen in Nebraska was collected by me in 1889, in a creek near Wahoo [Saunders Co.]. It had comparatively short outer bracts, broadly oblong rays, about 1 inch long, [and] thicker, nearly entire, well connate leaves, and more glaucous aspect. I would, maybe, have taken No. 191 for a hybrid of the two if they had been growing together, but I did not see any other *Bidens* than this and *B. frondosa* in western Nebraska. Upper Lawrence Fork, Kimball Co., August 10 [NEB 188737]; Lodgepole, Cheyenne Co., August 19.

192. ***Thelesperma filifolium*** Gray [*T. f.* var. *intermedium* (Rydb.) Shinners] – Collected in Nebraska for the first time by me in 1890 [NEB 207891, Banner Co., see second note below]. The root is annual, biennial, seemingly also perennial; but there is no creeping rootstock. The divisions of the leaves are sometimes broader than the rachis. Prairies: Deuel Co., June 25; Rush Creek [Garden Co.], July 2 [US 61114, 76488, **isotypes** of *Thelesperma intermedium* Rydb.]; Sidney, Cheyenne Co., August 19 [NEB 207824].

NEB 139976 and 139980 are Rydberg's July 11, 1891, collections of this species from hills between Lawrence Fork and Pumpkin Seed Valley, Banner Co.; he labeled them "193" and cited them with 193 (below), not 192. Both specimens were annotated as *T. intermedium* Rydb. by L. Shinners in 1949.

A year earlier–1890–Rydberg collected this plant in Banner Co., as *T. tenuifolium, s.n.*, NEB 207891 and the **holotype** of *Thelesperma intermedium* Rydb. as *T. filifolia* (*Rydberg 192*, NY 273827). Rydberg in Bull. Torr. Bot. Cl. 27: 631-632 gives 1890 and Banner County, but US 61114 and 76488 show Deuel and 1891, apparently from reading the labels.

193. ***Thelesperma gracilis*** Gray [*T. megapotamicum* (Spreng.) Kuntze] – Common in western Nebraska.–McCuligan Canyon, Deuel [now Garden] Co., June 26 [NEB 140013]; hills in Banner Co., July 11 [see note with 192, above].

In July, 1890, Rydberg collected this species in Deuel Co. (NEB 140011, 207926).

194. ***Hymenopappus tenuifolius*** Pursh. var. . . . . .*sic* – I believe that this belongs to *H. tenuifolius*, although the corolla is light yellow and not white. I would have named it *H. flavescens* Gray, but for the range and for the achenes, which are long villose. Sandhills of Phelps Co., June 18 [NEB 189832].

195. ***Hymenopappus*** sp. . . . . .*sic* [*H. filifolius* Hook. var. *polycephalus* (Osterh.) B. L. Turner]. – I do not know where to refer this, the most common *Hymenopappus* of western Nebraska. It seems to be between *H. tenuifolius* and *H. filifolius*. The stems are generally several, clustered from a perennial caudex, slender, 12–15 in. high, with few stem-leaves, but not scapiform. The heads are small, hardly 4 lines; the bracts with yellowish tips; the corollas are generally bright yellow. Hills and canyons: between Lawrence Fork and Pumpkin Seed Valley, Banner Co., July 6 [NEB 189854]; Wild Cat Mountains, July 16 [NEB 189819]; maybe it is *H. luteus* of Nuttall.

In 1890, Rydberg collected this species in July in Deuel Co. (NEB 189787) and in August in Garden Co. (NEB 207893).

196. ***Actinella acaulis*** Nutt. [*Tetraneuris acaulis* (Pursh) Greene] – In the SYNOPTICAL FLORA, and also M. Coulter's MANUAL, this species is said to have 3–5 in. long rays, which is undoubtedly a misprint not corrected in the reprint.–Hills of the west, common. McCuligan Canyon, Deuel [now Garden] Co., June 26; Rush Creek [Garden Co.], July 2; Scotts Bluff, July 24 [NEB 190636].

In 1890, Rydberg collected this species in Banner Co. (NEB 190639) and from McCuligan canyon in today's Garden Co. (NEB 207925).

197. ***Helenium autumnale*** L. – Platte bottoms, Scotts Bluff Co., July 24 [NEB 189765, 207823] [NEB 189751, 190681 are Rydberg, s.n., Fort Kearney, Kearney Co., September 1891.]

198. ***Bahia oppositifolia*** Nutt. [*Picradeniopsis oppositifolia* (Nutt.) Rydb. ex Britt.] – Salt marsh near the Wyoming line, [Scotts Bluff Co., according to label] July 31 [NEB 190681].

199. ***Chaenactis douglasii*** Hook. & Arn. var. ***alpina*** Gray (??) [*sic*] – As there are no flowers, and the fruits found consist of 3 incomplete achenes, I am in doubt if the determination is right. There are no specimens in the Herbarium [NEB] of the University. The general appeance, the receptacle and the involucre seems to indicate that the plant belonged to HELENIOIDEAE, and the involute bracts made me suppose that it belongs to *Chaenactis*. The plant is new to Nebraska [NEB 188970, 191488*. It was first noticed by Mr. J. Flodman on the hills of Upper Lawrence Fork**, Kimball Co., August 10

J. M. Holzinger: "More material for complete determination is a desideratum."

* In 1926, Rydberg wrote "*Senecio lanatifolius* Osterh. P.A.R." above his handwritten label on NEB 191488, but T. Barkley wrote "No!" near that in the 1970s.

** The site is more likely to be the Wildcat Hills of adjacent Banner Co., and both sites would be easternmost for the species. The plant has not been collected again in Nebraska.

200. ***Dys[s]odia chrysanthemoides*** Lag. [*Dyssodia papposa* (Vent.) A. S. Hitchc.] – Throughout Nebraska, common. Prairie, near Horse Creek, Scotts Bluff Co., August 1 [NEB 189182].

201. ***Pectis angustifolia*** Torr. – Collected for the first time in Nebraska, in 1890 [NEB 190662, Deuel Co., July, 1890, *s.n.*]. A sand-draw in Deuel Co., June 27 and August 24 [NEB 190660, 190661, 207825].

Rydberg also collected this species in the same place in July, 1890 (NEB 190662, 207931). It is common there and in adjacent counties.

202. ***Achillea millefolium*** L. – Prairie, Kiowa Valley [Scotts Bluff Co.], June 23 [NEB 199377, *Rydberg 202*, is from July 8 at this site].

203. ***Artemisia canadensis*** Michx. [*A. campestris* L.] – Hills of western Nebraska: Kiowa Valley, Scotts Bluff Co., July 28; Potter, Cheyenne Co., August 15 [NEB 207827]; sand-draw in Deuel Co., August 24 [NEB 188448].

Rydberg also collected this species in Banner Co. in August, 1890 (NEB 188352).

204. ***Artemisia dracunculoides*** Pursh [*A. dracunculus* L.] – Throughout Nebraska.– McCuligan Canyon, Deuel [now Garden] Co., August 22 [NEB 188426].

205. ***Artemisia biennis*** Willd. – Sand, Potter, Cheyenne Co., August 15.

206. ***Artemisia filifolia*** Torr. – Sandy soil in western Nebraska.–McCuligan Canyon, Deuel [now Garden] Co., August 22 [NEB 188454].

Rydberg also collected this species in modern Deuel or Garden Co. in July, 1890 (NEB 188453, 188455).

207. ***Artemisia frigida*** Willd. – Hills, Lawrence Fork, Banner Co., July 10 [NEB 188482].

Rydberg also collected this species in Banner Co., August, 1890 (NEB 188491).

208. ***Artemisia ludoviciana*** Nutt. – Throughout Nebraska, common.–Hills: Eagle Nest, Scotts Bluff Co., July 28; south of Kimball, August 13 [NEB 188555, 207817]; 208b is a slender variety with smaller leaves; not in bloom, although collected at the same time as the other form, August 13 [NEB 188517].

209. ***Artemisia cana*** Pursh – Not in bloom yet. Canyon in Kiowa Valley, Scotts Bluff Co., July 25 [NEB 188359, 188360].

210. ***Senecio canus*** Hook. var. . . . . . *sic* [*Packera cana* (Hook.) W. A. Weber & Á. Löve] – Tomentose throughout, 6–12 in. high; heads several; leaves nearly always entire, the radical oval or broadly oblong, petioled, sometimes with the petiole fully 4 in.; achenes glabrous or slightly and shortly strigose on the angles; involucre floccose or glabrous. It is undoubtedly nearly related to *S. canus* and *S. verneriaefolius*. Hills, south of Pumpkin Seed Valley, Banner Co., July 15 [NEB 191415]; Scotts Bluff, July 24.

J. M. Holzinger: "It is only the species."

211. ***Senecio aureus*** L. var. . . . . . *sic* [*Packera plattensis* (Nutt.) Weber & Á. Löve] – Platte [River], Kearney Co., June 15 [NEB 191501]; Cheyenne Co., July 3. [NEB 207818 is *Rydberg 211*, August 3, N. Platte River bottom, Morrill Co.]

J. M. Holzinger: "It is only the species."

212. ***Senecio douglasii*** DC. [S. riddellii Torr. & A. Gray] – Valleys: Lawrence Fork, Banner Co., July 10 [NEB 191589].

213. ***Cnicus pitcheri*** Torr. [*Cirsium undulatum* (Nutt.) Spreng.] – Prairies: Kearney Co., June 25; hills of Banner Co., July 6 [NEB 188919]; Scotts Bluff, July 24.

214. ***Cnicus ochrocentrus*** Gray [*Cirsium ochrocentrum* A. Gray] – Prairie, Kimball, August 12 [NEB 188944, 188945].

215. ***Cnicus undulatus*** Gray [*Cirsium undulatum* (Nutt.) Spreng.] – I would have taken the preceding for a variety of this, had it not been blooming about a month later. Hills between Lawrence Fork and Pumpkin Seed Valley, July 26; Curtis, Frontier Co., June 22; Deuel Co., July 25 [NEB 188923].

NY 24000 is *Rydberg 215, Platanthera praeclara* Sheviak & Bowles, June 15, 1891, Kearney County. That plant is in Rydberg's *Report* for that day but is not in this *List* under any name or number, and there is no voucher at NEB.

216. ***Cnicus altissimus*** Willd. var. ***filipendulum*** Gray [*Cirsium flodmanii* (Rydb.) Arthur] – Platte Valley, near Horse Creek, Scotts Bluff Co., August 2 [NEB 198233].

217. ***Crepis runcinata*** T. & G. – Platte bottoms, Cheyenne Co., July 3 [NEB 187257, 207819]; Pumpkin Seed Valley, Banner Co., July 14 [NEB 187251].

218. ***Lactuca ludoviciana*** DC. – Lawrence Fork [Banner Co.], July 10 [NEB 187321]; Mitchell Valley, Scotts Bluff Co., August 4.

219. ***Stephanomeria minor*** Nutt. [*S. runcinata* Nutt.] – I would have referred [this] to *S. runcinata* Nutt., but the pappus is plumose to the base. Badlands north of Scotts Bluff, July 22. It is easily taken for a *Lygodesmia*. [NEB 187576, 187577, 187666, 207822, **isotypes** of *Ptiloria ramosa* Rydb.; NY 114892, 114893, **isotypes**; RM 97231s; US 75555, **isotype]**

"219" is *Cucurbita foetidissima* in Rydberg's *Report*.

220. ***Lygodesmia juncea*** Don [*L. juncea* (Pursh) D. Don ex Hook.] – Unusually large specimens from a canyon north of Freeport, Banner Co., July 15 [NEB 187446].

221. See page 11. [*sic*]

221 is neither mentioned on that page of the original manuscript, nor is any other Asteraceae.

LOBELIACEAE

222. ***Lobelia spicata*** Lam. – [North] Platte bottoms, Scotts Bluff Co., August 25 *sic*

[August 3*] [NEB 187121, 207443, 207844 and US 57811 are from this site on August 3.]

*On August 25, Rydberg was in Deuel Co., 100 miles away, and the date he wrote in his manuscript is incorrect. The labels on the specimens note August 3, in his hand.

Rydberg collected *L. siphilitica* L. in Banner Co., July, 1890 (NEB 207920), the only such collection at NEB from that county.

CAMPANULACEAE

223. ***Campanula rotundifolia*** L. – Carter Canyon, Scotts Bluff Co., July 25 [NEB 187017, 187033, 207842].

PRIMULACEAE

224. ***Dodecantheon*** *sic* ***meadia*** L. [*Dodecatheon pulchellum* (Raf.) Merrill] – Only in fruit. [North] Platte bottoms, Cheyenne [now Morrill] Co., July 3.

Dodecatheon pulchellum is locally common in the North Platte River bottoms in Morrill and nearby counties. Neither the North nor South Platte rivers are in modern Cheyenne Co., but in 1891 that then-larger county included what is now Morrill Co. *Dodecatheon meadia* is unknown in Nebraska (Kaul et al., 2011, FLORA OF NEBRASKA).

OLEACEAE

225. ***Fraxinus pubescens*** Lam. [*F. pennsylvanica* Marsh. *s.l.*] – Rare in Nebraska, especially in the western part of the state. We saw a few trees in a gorge, on Little Medicine Creek, near Curtis, Frontier Co., June 22 [NEB 207841], and also a few at Hackberry Springs, Banner Co., July 10 [NEB 182320]. One at the latter place measured over 7 feet in circumference.

Fraxinus viridis Michx. *f.* [*F. pennsylvanica* Marsh.] is common throughout Nebraska. It was seen at several places, but no specimens were collected.

APOCYNACEAE

226. ***Apocynum cannabinum*** L. – Hills in Kearney Co., June 13 [NEB 182586]; Scotts Bluff, July 22.

ASCLEPIADACEAE

227. ***Asclepias speciosa*** Torr. – Sand-draw, Deuel Co., June 25 [NEB 192751]; Lawrence Fork, Banner Co., July 8.

228. ***Asclepias arenaria*** Torr. – In buds: sandhills, Kearney Co., June 13 [NEB 207845]; sand-draw, Deuel Co., in fruit August 23.

229. ***Asclepias verticillata*** L. var. ***pumila*** Gray [*A. pumila* (A. Gray) Vail] – Sandy places. Lawrence Fork, Banner Co., July 8 [NEB 182859].

Rydberg also collected this species in Banner Co. in August, 1890 (NEB 207885).

230. ***Acerates auriculata*** Eng. [*Asclepias engelmanniana* Woodson] – It is very likely to be confounded with *Asclepias stenophylla* Gray, which grows on the sandhills of Kearney Co. The form of the hood, and the auricles and the absence of the horn in *Acerates auriculata* are the only characters that separate the two. It was collected by me for the first time in Nebraska in 1890 [*s.n.*, Deuel Co.: NEB 182660, 182714, 207924]. McCuligan Canyon, Deuel [now Garden] Co., June 26; Scotts Bluff, July 24; sand-draw, near Kimball, August 12; canyon in Banner Co., August 8 [NEB 182720, 182721].

Rydberg also collected this species in Deuel or present-day Garden County in July, 1890 (NEB 207924).

231. ***Acerates viridifolia*** *sic* Ell. [*Asclepias viridiflora* Raf.] – Hills north of Lawrence Fork, Banner Co., July 10 [NEB 182658].

Rydberg also collected this species in Deuel or Garden County in August, 1890 (Rydberg 231, NEB 182656).

232. ***Acerates viridifolia*** *sic* Ell. var. ***lanceolata*** Gray [*Asclepias viridiflora* Raf.] – Hills north of Lawrence Fork, Banner Co., July 10 [NEB 182633].

GENTIANACEAE

233. ***Erythraea douglasii*** Gray [*Centaurium exaltatum* (Sw.) Druce] – New to Nebraska. Sands of Platte River, near Horse Creek, August 1 [NEB 182787]; and west of Scotts Bluff, Aug. 3.

234. ***Eustoma russellianum*** Griseb. [*E. grandiflorum* (Raf.) Shinners] – Platte bottoms near Horse Creek, Scotts Bluff Co., August 1 [NEB 182424]; Lodgepole Creek, near Kimball, August 12.

POLEMONIACEAE

235. ***Phlox muscoides*** Nutt. [*P. bryoides* Nutt.] – No flowers were found, and only a few specimens in fruit. There are no specimens of this or *P. bryoides* in the Herbarium [NEB] of the University. I have never seen any of them, but the likeness of the capsule of this and of the next made me believe it to be a *Phlox*. I named it *P. muscoides*, although the leaves are plainly quadrifarious. It was first collected by Mr. J. H. Flodman. Hills of Upper Lawrence Fork, Kimball Co., August 10 [NEB 183397; NY 336979].

236. ***Phlox hoodii*** Richards. – It was seen in several places in Banner and Scotts Bluff counties, but out of bloom. The only specimens in flower were found on rocky hills in Pleasant Valley, Scotts Bluff Co., near the Wyoming line, on July 30 [NEB 183436]. It has also been reported from N.W. Nebraska [It is widespread there].

237. ***Phlox caespitosa*** Nutt. var. ***rigida*** Gray (?) *sic* [*P. hoodii* Rich.] – Plants were seen in the Wild Cat Mountains, but then taken for a glabrate form of the preceding until one specimen in bloom was collected on a hill south of Scotts Bluff, July 26, and another in Kiowa Valley, July 29. The leaves are without cobwebby hairs, acerose, 2–3" long, at last spreading; corolla blue or white; lobes obovate, entire; tube longer than the calyx, which is a little cobwebby. It is new to Nebraska.

J. M. Holzinger: "*Phlox hoodii*, = 236."

238. ***Phlox douglasii*** Hook. var. . . . . . *sic* [*Phlox andicola* E. E. Nelson] – I refer this to *P. douglasii* Hook., which it resembles. The leaves, stem, and calyx are, however, very plainly cobwebby, which would associate it with *P. hoodii* and *P. canescens*. It certainly belongs to neither of them. Prairies and hills near Scotts Bluff, July 24 [NEB 207840]; hills of Upper Lawrence Fork, Kimball Co., August 10 [NEB 183387, 207839].

J. M. Holzinger: "It is the species."

239. ***Phlox douglasii*** Hook. var. ***longifolia*** Gray [*P. andicola* E. E. Nelson] – Valley near Rush Creek, Deuel Co., July 2.

J. M. Holzinger: "Phlox bryoides" [Numbered 235 in Holzinger's list, but between 238 and 246.]

240. ***Gilia linearis*** (Nutt.) Gray [*Collomia linearis* Nutt.] – New to Nebraska. Sand-draw in Deuel Co., June 25 [NEB 183305, 207837].

241. ***Gilia gracilis*** Hook. [*Microsteris gracilis* (Hook.) Greene var. *humilior* (Hook) Cronq.] – Also new to the state. In the same locality as the preceding, June 25 [NEB 183356, 183358].

242. ***Gilia longiflora*** Don [*Ipomopsis longiflora* (Torr.) V. E. Grant] – Sandhills of Kearney Co., August 29 [NEB 183335, 183345].

243. ***Gilia pungens*** Benth. var. ***caespitosa*** Gray [*Leptodactylon caespitosum* Nutt.] – It is given in Coulter's MANUAL for Scotts Bluff, Wyoming *sic*. I found it common on Wild Cat Mountains and on the hills of Scotts Bluff Co., especially on Scotts Bluff, Nebraska. It is not given in Webber's CATALOGUE. Scotts Bluff, July 25 [NEB 183285]; hills in Kiowa Valley, Scotts Bluff Co., July 29 [NEB 183286; also

mentioned in the *Report* as collected on "hills in Pleasant Valley near the Wyoming line."]

244. ***Gilia spicata*** Nutt. [*Ipomopsis spicata* (Nutt.) V. E. Grant] – New to Nebraska. There are no specimens of this in the Herbarium [NEB] of the University of Nebraska. Hills north of Lawrence Fork, Banner Co., July 6 [NEB 183354, 207835] and 10.

245. ***Gila iberidifolia*** Benth. [*Ipomopsis congesta* (Hook.) V. E. Grant] – Cliffs and canyons of Banner and Scotts Bluff Co. Wild Cat Mountains, Banner Co., July 16; Scotts Bluff, July 26 [NEB 183245, 183284, 183293].

246. ***Gilia*** sp. . . . . .[G. *pinnatifida* Nutt. ex A. Gray] – Glandular pubescent; root biennial or maybe perennial; stem branched, about 1 foot high or higher; leaves somewhat irregularly pinnatifid; lobes linear, not wider than the rachis, somewhat fleshy, mucronate; flowers in a branched panicle; corolla violet or blue with a white tube, somewhat funnelform, 2 lines long; stamens exerted; calyx with prominent ribs. It may be a variety of *G. pinnatifida*, which I have never seen. The corolla, however, is smaller than in the description and not strictly salverform. It is new to the state. Sands of [North] Platte River near Horse Creek, Scotts Bluff Co., August 1 [NEB 183351].

J. M. Holzinger: "*Gilia pinnatifida*."

HYDROPHYLLACEAE

247. ***Phacelia*** sp. . . . . . .[*Phacelia hastata* Dougl. ex Lehm.] – It was collected for the first time in Nebraska in 1890, by Mr. T. A. Williams, in the foothills [in Nebraska] of Black Hills. This year, I found a new station at Scotts Bluff, July 22 [NEB 183511, 183555, 207886; RM 238975]. I would have named it *P. circinata* Jacq., the description which agrees very well, but is not at all like *P. circinata* of C. C. Perry's collection of 1876 from North California, which is strigose, not canescent, has shorter and smaller flowers, nearly glabrous filaments, and the leaves variously lobed. In all plants I have collected the leaves are entire, but in one specimen of Mr. T. A. Williams', there were a few small pinnae below. The leaves are silky canescent on both sides, strigose only on the pedicels; flowers light blue or whitish; root-leaves spatulate-oblanceolate, entire; filaments bearded.

J. M. Holzinger: "*Phacelia circinata* Jacq."

BORAGINACEAE

248. ***Heliotropium currasivicum*** *sic* L. [*H. curassavicum* L. var. *obovatum* DC.] – Wet meadow, saline soil in Pumpkin Seed Valley, Banner Co., July 16 [NEB 183651].

249. ***Echinospermum floribundum*** Lehm. [*Hackelia floribunda* (Lehm.) I. M. Johnst.] – It is not given in Webber's CATALOGUE, but was collected in 1890 by Mr. T. A. Williams in War Bonnet Canyon in [Sioux Co.] N. W. Nebraska. A new locality was found this summer, July 9 [NEB 183679], in the plum-thickets in Lawrence Fork, Banner Co.

250. ***Echinospermum redowskii*** Lehm. var. ***occidentale*** Watson [*Lappula redowskii* (Hornem.) Greene var. *redowskii*] – Dooryard, Phelps Co., June 18 [NEB 183700].

Rydberg also collected this species in Deuel or present-day Garden Co., August, 1890 (NEB 183694).

251. ***Echinospermum redowskii*** Lehm. var. ***cupulatum*** Gray [*Lappula redowskii* (Hornem.) Greene var. *cupulata* (A. Gray) M. E. Jones] – Prairie, Venango, Perkins Co., June 23 [NEB 183722, 207847].

252. ***Krynitzkia crassisepala*** Gray [*Cryptantha minima* Rydb.] – Sandy prairies, especially where the sod had been disturbed, as in prairie-dog towns. Deuel Co., June 25 [NEB 183616, 183621].

253. ***Krynitzkia fendleri*** Gray [*Cryptantha fendleri* (A. Gray) Greene] – Wild Cat Mts., Banner Co., July 17 [NEB 183598, 183612]. It is not given in Webber's CATALOGUE, but was

afterward collected in 1890 [in Sioux Co.] by Rev. Bates of Valentine, Neb.

254. ***Krynitzkia jamesii*** Gray [*Cryptantha cinerea* (Greene) Cronq. var. *jamesii* (Torr.) Cronq.] – Not given in Webber's CATALOGUE. It was collected by me in 1890 [Banner Co., NEB 183971, 183988, 183989, 207923]. Sandy prairies, Deuel Co., June 25 [NEB 183977, 207847]; Wild Cat Mts., July 16; Lawrence Fork, Banner Co., July 8

255. ***Krynitzkia glomerata*** Gray [*Cryptantha thyrsiflora* (Greene) Payson] – Rocky hills: McCuligan Canyon, Deuel [now Garden] Co., June 26 [NEB 207848]; Rush Creek, Deuel [now Garden] Co., July 2 [NEB 114094, 184004; RM 82748s]; Lawrence Fork, Banner Co., July 9 [NEB 207849]; Scotts Bluff, July 24.

256. ***Krynitzkia sericea*** Gray (?) *sic* [*Cryptantha cana* (A. Nelson) Payson] – As no specimens in flower were seen, it might, perhaps, be *K. fulvocanescens*. Neither of the two are given in Webber's CATALOGUE.– Cliffs: McCuligan Canyon, Deuel [now Garden] Co., June 26 [NEB 183944]; Upper Lawrence Fork, Kimball C., August 10.

257. ***Mertensia lanceolata*** DC. – In fruit, among bushes in Lawrence Fork, Banner Co., July 3.

258. ***Lithospermum hirtum*** Lehm. [*L. carolinense*?] – Sandhills of Kearney Co., June 15 [possibly 13].

259. ***Lithospermum angustifolium*** Michx. var. . . . .*sic* [*L. incisum* Lehm.] – It was collected only in fruit. It is very low and much branched, but is, undoubtedly, nothing but a stunted form of *L. angustifolium*.–Prairies, Venango, Perkins Co., June 23 [NEB 183873].

260. ***Onosmodium carolinianum*** DC. [*O. molle* var. *occidentale* Mack.] – Prairies, Kearney Co., June 13. [NY 335611, *Rydberg s.n.* at same site and day, **lectotype** of *O. occidentale* Mack.].

CONVOLVULACEAE

261. ***Ipomaea*** *sic* ***leptophylla*** Torr. – Prairies of western Nebraska. Lawrence Fork, Banner Co., July 8 [NEB 183087, 207850].

In July, 1890, Rydberg collected this species from sand draws in Deuel (or present-day Garden) Co. (NEB 183080).

262. ***Convolvulus sepium*** L. [*Calystegia sepium* (L.) R. Br.] – Roadside east of Potter, Cheyenne Co., Aug. 17.

263. ***Evolvulus argenteus*** Pursh [*E. nuttallianus* J. A. Schultes] – Sandhills, Kearney Co., June 15 [NEB 183041]; sand-draw, Deuel Co., June 25.

Rydberg also collected this species in Deuel Co., July, 1890 (NEB 183035, 207921).

264. ***Cuscuta*** sp. . . . . . *sic* – I do not know to which this should be referred. I at first took it for *C. gronovii* Willd., but the capsule is depressed globular, which would associate it with *C. arvensis* and *C. tenuiflora*. It differs from the former therein that the flower-clusters are more or less cymose, not dense, and from the latter in being more slender and in having a broadly campanulate corolla with spreading lobes. More slender specimens resemble somewhat *C. californica*, but the lobes of the corolla are obtuse.–On *Chenopodium*, *Polygonum*, and occasionally on *Helianthus annus (sic)* and *Psoralea lanceolata*.–Canyon in Kiowa Valley, Scotts Bluff Co., July 29.

J. M. Holzinger: "*Cuscuta indecora*."

Rydberg collected *C. indecora* in Banner Co., August, 1890 (NEB 183166, 183174, 207922).

SOLANACEAE

265. ***Solanum triflorum*** Nutt. – This, *Krynitzkia crassisepala*, and *Munroa squarrosa* frequent the prairie dog towns. Lawrence Fork, Banner Co., July 8 [NEB 18546_ (final digit illegible), 207851; NY 820284, 820258].

266. ***Solanum nigrum*** L. [*S. interius* Rydb.] – McCuligan Canyon, Deuel [now Garden] Co., August 22.

267. ***Solanum rostratum*** Dunal – It is becoming more and more common eastward, and in some places even a troublesome weed. Broken soil, Deuel [now Garden] Co., July 1. [NEB 185452, 207853; NY 820854, 820855].

> Labels on NEB specimens say only "broken soil," in Rydberg's hand, but those of NY specimens say "Rush Creek". That creek is in modern Garden County, which was carved out of then-larger Deuel Co. in 1908.

268. ***Physalis virginiana*** Mill. [*P. heterophylla* Nees] – Railroad banks, Kearney Co., June 15 [NEB 185177], and Curtis, Frontier Co., June 22 [NEB 185165].

269. ***Physalis mollis*** Nutt. var. ***cinerascens*** Gray [*P. hederifolia* A. Gray var. *comata* (Rydb.) Waterfall, *P. comata* Rydb.] – The species of *Physalis* are very perplexing, especially those of western Nebraska. In eastern Nebraska, I have collected 3, viz. *P. angulata, virginiana*, and *lanceolata*. All agree well with the descriptions in the SYNOPTICAL FLORA. *P. lanceolata* var. *laevigata* Gray is also given in Webber's CATALOGUE, but I have not seen any specimens from eastern Nebraska. In western Nebraska, I have collected 5 different forms (species or varieties), which have given me more or less trouble. In fact, I do not know to which to refer them. The first I have referred to *P. mollis* Nutt. var. *cinerascens* Gray. It is very similar to the specimens in the Herbarium [NEB] of the University. The leaves are broadly ovate, 2 in. long, angulately toothed or repand, very thin; pubescence stellate or simple (even a little glandular), very short, except on the calyx, the pedicels, the petioles and the younger parts of the stem, where it is long and soft; petioles as long as the leaves or longer; corolla greenish yellow with darker roots; anthers yellow; fruiting calyx globose ovate.–Not collected in the state before. Under the cliffs on the south side of Scotts Bluff, July 20 [NEB 185153 **isotype** of *P. comata* Rydb.; NY 138822 **holotype**; NY 138823 **isotype**; US 126246 **isotype**].

270. ***Physalis*** sp. . . . .*sic* [*P. hederifolia* A. Gray var. *comata* (Rydb.) Waterfall] – It is very like the preceding and likely only a variety of the same species as it. The leaves, however, are smaller, scarcely over 1 inch long, thickish; petioles generally shorter than the blades; flowers not seen; fruiting calyx globose, about 1 inch. It is more erect than the preceding; the pubescence is shorter, and the long hairs very few. The general appearance is like that of *P. fendleri* in the Herbarium of the University of Nebraska [NEB], but the pubescence of my specimens is not microscopically minute. Prairie near Ashford, Banner Co., August 8 [NEB 185282].

> J. M. Holzinger: "*Physalis*–Nearest to *hederaefolia*, but out of range for this.—I would do no better with this."

271. ***Physalis*** sp. . . . . . [*P. longifolia* Nutt.] – This is also very likely only a variety of the same species as the two preceding. The general habit is between the two. The leaves are thicker and shorter petioled than in No. 270, and longer than in No . 271. Fruiting calyx ovate, about 1 inch; stem low and nearly prostrate. The pubescence is more glandular, and the strong scent is perceived at a distance of several rods. It seems to be a connecting form between the "Mollis" group and *P. palmeri*. Sandy banks of a draw south of Sidney, Cheyenne Co., August 18 [NEB 185231, 185267].

> J. M. Holzinger: "Physalis–palmeri? May equal [Rydberg's] 270."

272. ***Physalis longifolia*** Nutt. – I collected this and the following last year*, and sent some specimens to Dr. Watson, who said that both belong to *P. lanceolata*, and that this is the variety *laevigata* Gray. For my part, I believe that it has full right to a specific rank, as I have not seen any intermediate forms. If this is *P. lanceolata* var. *laevigata* Gray, I would add to the characters in the SYNOPTICAL

FLORA: berry *yellow*, stalked in the calyx; this ovate, less angled, and *not sunken* at the base; bottoms of the calyx, stipe, and lower part of the berry very *glutinous*, as if they were *painted with molasses*. It is these characters that make me believe it to be distinct, hence I named it with Nuttall's name, *P. longifolia*. Lawrence Fork, Banner Co., July 8 [NEB 185269]; near Lodgepole Creek, Kimball, August 12 [*P. hispida* (Waterfall) Cronq., NEB 185254, 207854]

*Deuel Co., July, 1890 (NEB 185265).

273. ***Physalis*** sp. . . . . *sic* [*P. hispida* (Waterfall) Cronquist] – It belongs to the LANCEOLATA group and stands very near to *P. lanceolata* and the preceding. However, it agrees better with the description of *P. angustifolia* than to that of *P. lanceolata*, except in the pubescence, which is short and stiff or, more commonly, scarcely any, except on the calyx. The plant is low, from 6–10 in., prostrate to nearly upright; leaves obovate, nearly always entire; fruiting calyx globose ovate, scarcely angled, and very little sunken at the base; berry yellowish green; corolla ½ in. wide, greenish yellow with a darker spot (in *P. lanceolata* it is 2/3–3/4 in wide and ochroleucous). Sand-draw, Deuel Co., June 25; Lawrence Fork, Banner Co., July 8 [NEB 185234; RM 2503]; Rush Creek, Deuel [now Garden] Co., July 2; Scotts Bluff, July 24 [NEB 185243]; Sidney, Cheyenne Co., August 17.

J. M. Holzinger: "*Physalis viscosa* var. *spathulaefolia* Gr., apparently a prostrate form. Some of your plants under this no. were *Physalis lanceolata*."

274. ***Physalis lanceolata*** Michx. [*P. virginiana* P. Mill.] – Railroad bank, Kearney Co., Jun. 13 [15 on label] [NEB 185326].

SCROPHULARIACEAE

275. ***Penstemon frimontii*** *sic* T. & G. var. . . . (?) *sic* [*P. albidus* Nutt.] – *Fremontii* is not given for Nebraska, and no specimens of the same are in the Herbarium [NEB] of the University. Prairie, Deuel Co., June 27 [NY 541697].

J. M. Holzinger: "*Penstemon acuminatum*."

276. ***Penstemon glaber*** Pursh [var. *glaber*] – Common west: Lawrence Fork, Banner Co., July 8. [NEB 185933, 185934; NY 595804].

277. ***Penstemon grandiflorus***. – Prairie, Kearney Co., June 13 [NY 601122; RM 2515].

278. ***Penstemon humilis*** Nutt. (?) *sic* [*P. albidus* Nutt.] – Corolla blue or white.–Dr. Hapeman of Minden called it *P. albidus*, the description of which fits very well, except that the corolla is plainly bilabiate. There are no specimens in the Herbarium [NEB] of the University. Prairies: Kearney Co., June 15 [NEB 185847]; Gosper Co., June 20 [NY 541698]; Deuel Co., June 25 [NY 541697].

J. M. Holzinger: "*Penstemon albidus*".

279. ***Penstemon gracilis*** Nutt. [var. *gracilis*] – Wet meadow, Deuel Co., June 26 [NEB 186060, 207855; NY 600776, 600850].

280. ***Mimulus jamesii*** T & G. [*M. glabratus* Kunth var. *jamesii* (Torr. & A. Gray ex Kunth) A. Gray] – In brooks: Lawrence Fork, Banner Co., July 10; Horse Creek, Scotts Bluff Co., August 1 [NEB 185697].

281. ***Limosella aquatica*** L. – Spring in Deuel Co., June 27 [NEB 185655].

282. ***Veronica americana*** Schwein. – In water, Lawrence Fork, Banner Co. [July 10] [NEB 186140, 186146; RM 2517].

283. ***Veronica peregrina*** L. var. . . . .*sic* [var. *xalapensis* (Kunth) Pennell] – The plant is glandular and the pod obcordate, deeply notched, but the general appearance is that of *V. peregrina*, not that of *V. arvensis*. Only a few specimens were collected on a hill near Kimball, August 12.

284. ***Gerardia purpurea*** L. [*Agalinis purpurea* Pennell] – Island in Platte River, near Kearney City [Buffalo or Kearney County] August 29.

The river separates Buffalo County from Kearney County, and either is the possible collection site. Kearney City is in Buffalo Co., not Kearney Co.

285. ***Gerardia tenuifolia*** Vahl. [*Agalinis tenuifolia* (M. Vahl) Raf. var. *parviflora* (Britt.) Britt.] – Common in western Nebraska, in low places: Lawrence Fork, Banner Co., July 10; Lodgepole, east of Potter, Cheyenne Co., August 18.

Rydberg also collected this species in Banner Co. in August, 1890 (NEB 185564, 207919).

286. ***Castileia*** *sic* ***sessiliflora*** Pursh – It has been collected in the Black Hill region by T. A. Williams, in 1890. Hills of Deuel Co., June 25 [NEB 185601].

287. ***Orthocarpus luteus*** Nutt. – Lodgepole Creek, near Kimball, August 12 [NEB 120251, 185809, 185811].

288. ***Aphyllon fasciculatum*** Gray [*Orobanche fasciculata* Nutt.] – Sandhills, Kearney Co., June 13; hills, Deuel Co., June 27; Rush Creek, Deuel [now Garden] Co., July 2.

289. ***Aphyllon fasciculatum*** Gray. var. ***luteum*** Gray [*Orobanche fasciculata* Nutt.] – It was collected for the first time in Nebraska in 1890, by Rev. Bates of Valentine. Only a few specimens were collected [by us] on the hills south of Scotts Bluff, July 26 [NEB 186204].

290. ***Aphyllon ludovicianum*** Gray [*Orobanche ludoviciana* Nutt.] – It is undoubtedly *A. ludovicianum*, although the calyx is almost 5-parted. The flowers are yellowish, with purplish limb, and plainly 2-lipped. If parasitic at all, on the roots of *Crysopsis*, in sand-draws: Lawrence Fork, Banner Co., June 10 [NEB 186206]; 290b is a depauperate form from the hills of Upper Lawrence Fork, Kimball Co. August 10.

Rydberg also collected this species in Deuel Co., July, 1890 (NEB 122177).

291. ***Aphyllon ludovicianum*** Gray . . . . *.sic* [*Orobanche ludoviciana* Nutt.] – Plant dark brown, flowers dark purple. – On the root of *Opuntia missouriensis* (?) *sic*. Prairies, north of Kimball, August 12.

J. M. Holzinger: "There was so little of this number that I simply had the plant placed in the Herbarium [US] with your suggestion. The material was too scant for an investigation. This seems to be a more slender plant than the species."

LABIATAE [LAMIACEAE]

292. ***Teucrium occidentale*** Gray [*T. canadense* L. var. *occidentale* (A. Gray) McClint. & Epling] – Spring near Horse Creek, Scotts Bluff Co., August 1.

293. ***Mentha canadensis*** L.– [*M. arvensis* L.] Lawrence Fork, Banner Co., July 11 *sic* [10 on label]; Horse Creek, Scotts Bluff Co., August 1 [NEB 184607].

Rydberg also collected this species at Hackberry Springs, Banner Co., in August, 1890 (NEB 184605).

294. ***Lycopus lucidus*** Turcz. var. ***americanus*** Gray [*Lycopus americanus* Muhl. ex W. Barton] – Pumpkin Seed Creek, Banner Co., July 18.

295. ***Lycopus sinuatus*** Ell. [*L. americanus* Muhl. ex W. Barton] – In Nebraska it is often stoloniferous.–Lawrence Fork, Banner Co., July 18.

Rydberg also collected this species in Banner Co. in 1890 (NEB 184413).

296. ***Hedeoma hispida*** Pursh – Sand-draw, Deuel Co., June 27.

297. ***Hedeoma drummondii*** Benth. – Common in western part of the state.–Court House

Rock, Cheyenne [now Morrill] Co., July 4; Scotts Bluff, July 24 [NEB 184350, 207857].

NEB 184340 is July 8, 1891, Lawrence Fork, Banner Co.

298. ***Salvia lanceolata*** Willd. [*S. reflexa* Hornem.] – Hill, Curtis, Frontier Co., June 22 [NEB 184821].

299. ***Monarda fistulosa*** L. [var. *menthifolia* (Graham) Fernald] – Canyon in Banner Co., July 9 [6, on label] [NEB 184656].

300. ***Monarda fistulosa*** L. var. ***mollis*** Benth. [var. *menthifolia* (Graham) Fernald] – Lawrence Fork, Banner Co., July 8 [NEB 184637, 184646]; Prairie in Kiowa Valley, Scotts Bluff Co., July 29 [NEB 184858].

301. ***Monarda citriodora*** Cerv. [*M. pectinata* Nutt.] – The specimens I collected do no fully agree with the description in the SYNOPTICAL FLORA. The flowers are white, except a spot on the lower lip, which is yellow and sprinkled with minute brown-purple spots. The bracts are greenish, not whitened nor purplish, but with strong ribs. Low prairies, Rush Creek, Deuel [now Garden] Co., July 2 [NEB 184653, 184712, 207858]; Lawrence Fork, Banner Co., July 8.

Rydberg also collected this species in Banner Co., August, 1890 (NEB 184709, 207918).

302. ***Scutellaria galericulata*** L.–Spring near Horse Creek, Scotts Bluff Co., August 1 [NEB 184902]. It is not given in Webber's CATALOGUE.

303. ***Physostegia parviflora*** Nutt. – Collected for the first time in the state. Spring near Horse Creek, Scotts Bluff Co., August 1.

PLANTAGINACEAE

304. ***Plantago*** sp. . . . . . [*P. eriopoda* Torr.] – It stands very near *P. eriopoda* Torr., has a mass of brownish hairs at the crown, but not so copious as in the named species. The ovules are one pair in each cell or sometimes more; leaves oval lanceolate–linear lanceolate, gradually tapering into a margined petiole, with this 4–8 in. long, rather thinnish; spike 3–18 in. long, on its upper half many flowered and sparse, in the manner of *P. rugellii.*; capsule, as in this, when well developed, twice the length of the calyx, circumscissile much below the middle. It was collected by Rev. Bates of Valentine, in 1890. He named it *P. lanceolata*, and remarked that he believed it to be indigenous. It certainly does not belong to *P. lanceolata*. If not a variety of *P. eriopoda*, it is very near to it. Platte bottoms, Kearney Co., June 15 [NEB 186369]; Pumpkin Seed Valley, Banner Co., July 19 [NEB 136370]. The first locality was shown to me by Dr. Hapeman of Minden.

J. M. Holzinger: "*Plantago eriopoda.*"

305. ***Plantago patagonica*** Jacq. var. ***gnaphalioides*** Gray [*P. patagonica* var. *patagonica*] – Hills, Kearney Co., June 15; prairie, near Venango, Perkins Co., June 23 [NEB 186427, 186435].

Rydberg also collected this species in Deuel Co., July, 1890 (NEB 186434).

VERBENACEAE

306. ***Lippia cuneifolia*** Steud. – The peduncles are generally a little longer than the leaves, and the stems ae sometimes rooting at the nodes. Sandy prairies: Phelps Co., June 18 *sic* [19 on label] [NEB 184027, 184048]; Kiowa Valley, Scotts Bluff Co., July 29.

NYCTAGINACEAE

307. ***Oxybaphus nyctaginea*** Sweet var. ***oblongifolius*** Gray [*Mirabilis hirsuta* (Pursh) MacMill.] – Hill, Banner Co., July 10 [NEB 172722].

308. ***Oxybaphus angustifolius*** Sweet [*Mirabilis linearis* (Pursh) Heimerl] – Hills: Scotts Bluff Co, July 24 [NEB 172678]; Kimball, August 12.

Rydberg also collected this species in Deuel Co. in July, 1890 (NEB 172677).

309. ***Abronia fragrans*** Nutt. – Valleys: Rush Creek, Deuel [now Garden] Co., July 2; Lawrence Fork, Banner Co., July 8 [NEB 172608].

J. M. Holzinger: "The most important omission [in your report] is a note with 339. With *Rumex venosus* I found one sheet of *Abronia micrantha* Torr. This is the more remarkable since you have done your work so very thoroughly, and this is new to Nebraska too."

In August, 1890, Rydberg collected this species in Pumpkin Creek Valley, Banner County (NEB 172590, 172597).

ILLECEBRACEAE [CARYOPHYLLACEAE]

310. ***Paronychia jamesii*** T. & G. – Barren hills: Deuel [now Garden] Co., June 25 [NEB 172942]; Scotts Bluff, July 24.

FICOIDEAE [MOLLUGINACEAE]

311. ***Mollugo verticillata*** L. – Sand-draw, Deuel Co., August 24.

AMARANTHACEAE

312. ***Amarant[h]us blitoides*** Watson [*A. albus* L.] – Cultivated soil: Rush Creek [Garden Co.], July 2; Kiowa Valley, Scotts Bluff Co., July 29 [NEB 171889].

313. ***Amarant[h]us*** sp. . . . . . .*sic* [*A. blitoides* S. Watson] – Stem prostrate, terete, and succulent; leaves as those of *A. blitoides*, but not as rigid; sepals and stamens 5. Maybe it belongs to *A. torreyi* Benth., which I have not seen. In a canyon east of Scotts Bluff Co., July 30, and on broken land in Pleasant Valley, Scotts Bluff Co., July 30 [NEB 171934, 171937]. At the latter station, it has become a somewhat troublesome weed. It has received the name "mountain purslane." In drying, the succulent *sic* becomes wrinkled and looks as furrowed as in *A. blitoides*, which it then resembles very much. New to Nebraska.

J. M. Holzinger: "The difference you point out seemed to us here probably [is] the result of difference in soil and nourishment. At best, it cannot be made a specific difference. And your 313 would be a variety of *Amaranthus blitoides*. But I doubt the feasibility of assigning it to even a var. *Chenopodium*."

314. ***Froelichia floridana*** Moq. – Sandhills, Kearney Co., August 29 [NEB 172049].

315. ***Froelichia gracilis*** Moq. – Leaves oblong lanceolate, mostly radical; calyx in fruit with a few sharp tubercles, or nearly even. Not reported for Nebraska. Sand-draw, Deuel Co., August 24 [NEB 172077, 172079].

CHENOPODIACEAE

316. ***Cycloloma platyphyllum*** Moq. [*C. atriplicifolium* (Speng.) J. M. Coult.] – Sandy soil: Near Court House Rock, Cheyenne [now Morrill] Co., July 3; Kiowa Valley, Scotts Bluff Co., July 30 [NEB 172429].

317. ***Chenopodium fremontii*** Watson var. ***incanum*** Wats. – Not in Webber's CATALOGUE. Sandy places: Lawrence Fork, Banner Co., July 8; Kiowa Valley, Scotts Bluff Co., August 1 [NEB 172312].

Rydberg also collected this in Banner Co., August, 1890 (NEB 172232, 172250).

318. ***Chenopodium leptophyllum*** Nutt. [*C. pratericola* Rydb.] –Common in Kearney Co., and in the west. A worse weed than *C. album*. Kearney Co., June 13 [NEB 172345, 172381].

319. ***Chenopodium leptophyllum*** Nutt. var. ***oblongifolium*** Wats. [*C. desiccatum* A. Nelson] – Broken soil, Deuel Co., July 2 [NEB 172347].

320. ***Chenopodium leptophyllum*** Nutt. var. ***subglabrum*** Wats. [*C. pratericola* Rydb.] –

This as well as the preceding is not in Webber's CATALOGUE. It seems to be very distinct from the true *C. leptophllyum*.–Hillsides: Scotts Bluff, July 24 [NEB 172362].

321. ***Chenopodium*** sp. . . . . . . [*C. fremontii* S. Watson] – I collected some undeveloped specimens of this last year, and sent some to Dr. Watson, remarking that it was much like the *C. opulifolium* of Europe, which is credited for North America in DeCandolle's PRODROMUS. Dr. Watson said that it was too undeveloped to identify, but that it probably was a slender variety of *C. album*. This year, I have seen it in several places, growing as well together with *album* as elsewhere. It always grows in shady and moist places. The only place I found it in fruit was in Carter Canyon, Scotts Bluff Co., July 26 [NEB 172248, 172258, 172259], and in McCuligan Canyon, Deuel [now Garden] Co., August 22. As I cannot find any description that fits it, I will here give a short one:

> Plant about 2 feet high, slender, branched, nearly destitute of mealiness, except on the calyx, which is very small and carinate; leaves very thin, 3-ribbed, 1½ – 2 in. long, broadly triangular, hastate-ovate lanceolate, entire or slightly toothed, mucronate; flowers in very small clusters on the slender branches; seeds easily separated from the pericarp.–Canyon in Banner Co., July 2 in bud.

J. M. Holzinger: "*Chenopodium fremontii*".

322. ***Chenopodium album*** L. [*C. berlandieri* Moq.] –Common. Carter Canyon, together with the last, July 26 [NEB 172173, 172190, 172198].

323. ***Monolepis chenopodioides*** Moq. [*M. nuttalliana* (Roem. & Schult.) Greene] – Dry saline soil, Deuel Co., June 27 [NEB 172559; NY 539850].

324. ***Atriplex patula*** L. var. ***hastata*** Gray [*A. dioica* Raf., *A. subspicata* (Nutt.) Rydb.] – Saline soil, near Horse Creek, Scotts Bluff Co., August 1 [NEB 172113].

325. ***Atriplex argentea*** Nutt. – It is not given in Webber's CATALOGUE, but mine was collected by me in 1890, in the salt basins near Lincoln [NEB 172093]. Saline soil in Kiowa Valley, Scotts Bluff Co., July 23 [NEB 172081, 172103].

325 [*sic*; 326 certainly intended] ***Atriplex nuttallii*** Watson [*A. canescens* (Pursh) Nutt.] – Canyon, east of Scotts Bluff, July 24 [NEB 172148, 172150; RM 234797]; Kiowa Valley, Scotts Bluff Co., July 31.

327. ***Eurotia lanata*** Moq. [*Krascheninnikovia lanata* (Pursh) A. Meeuse & A. Smit] – Hillsides: Lawrence Fork, Banner Co., July 3 [NEB 172488, 172491, 172494]; Scotts Bluff Co., July 26.

In 1890, Rydberg collected this species in Banner Co. in July (NEB 172479) and August (NEB 172480).

328. ***Copiospermum*** *sic* ***hyssopifolium*** L. [*Corispermum americanum* (Nutt.) Nutt.] – Canyon, east of Scotts Bluff, July 4. [NEB 172415]

329. ***Suaeda diffusa*** Watson [*S. calceoliformis* (Hook.) Moq.] – Saline meadows: Pumpkin Seed Valley, Banner Co., July 16 [NEB 172454]; Horse Creek, Scotts Bluff Co., August 1.

J. M. Holzinger: "*Suaeda linearis* var. *salsa*."

POLYGONACEAE

330. ***Eriogonum alatum*** Torr. [var. *alatum*] – Hills: McCuligan Canyon, Deuel [now Garden] Co., June 26 [NEB 171131; RM 2557]; Lawrence Fork, Banner Co., July 8; Big Horn, Banner Co., August 8.

In 1890, Rydberg got this species from Magnesia Cliffs, Garden Co. (NEB 171135), and from an unspecified site in Deuel or Garden Co. (NEB 171133).

331. ***Eriogonum flavum*** Nutt. [var. *flavum*] – Cliffs: McCuligan Canyon, Deuel [now Garden] Co., June 26 [NEB 171164]; Rush Creek [Garden Co.], July 2 [NEB 171154]; Upper Lawrence Fork, Kimball Co., August 10 [RM 237797.

332. ***Eriogonum annu[u]m*** *sic* Nutt. – Common in western Nebraska, on hills and prairies.– Lawrence Fork, Banner Co., July 8; Pumpkin Seed Valley, Banner Co., July 14; Kiowa Valley, Scotts Bluff Co., July 28 [NEB 171100]; Sidney, Cheyenne Co., August 18 [NEB 171088].

333. ***Eriogonum cernuum*** Nutt. – This species was collected for the first time in Nebraska, in 1890, by me on a hill in Banner Co. [NEB 171143], but the specimens were too immature for identification. This year, I have found several new localities. In fact, it is not uncommon in the west. Court House Rock, Cheyenne Co., July 4; Wild Cat Mts., Banner Co., July 16; Scotts Bluff, July 24 [NEB 171142]; Kimball, August 12; Sidney, Aug. 18.

Flodman also collected this species (NEB 171141) at Courthouse Rock on July 4, 1891. It is one of the few specimens from the 1891 trip that has his name on its label.

334. ***Eriogonum multiceps*** Nees [*E. pauciflorum* Pursh var. *gnaphalodes* (Benth.) Reveal] – Court House Rock, Cheyenne Co., July 4; Scotts Bluff, July 25 [NEB 171192]. It sometimes varies with long pedicels. Hills south of Scotts Bluff, August 4 (334b).

335. ***Eriogonum microthecum*** Nutt. var. . . . . . [*E. effusum* Nutt. var. *effusum*] – This, and also the next, differs from the specimens in the Herbarium [NEB] of the University of Nebraska, in that in my specimens the nerves of the involucre are barely seen, the leaves are broader, scarcely revolute, and scarcely glabrate on the upper side. Prairies in Kiowa Valley, Scotts Bluff Co., July 28 [NEB 171177].

J. M. Holzinger: "*Eriogonum microthecum* Nutt. approaching var. *fendleri*."

Rydberg collected this species in August, 1890, from sand draws in Cheyenne Co. (NEB 171144, 171145, 171176).

336. ***Eriogonum microthecum*** Nutt. var. ***effusum*** Gray & Torr. [*E. effusum* Nutt. var. *effusum*] – Leaves broader, not revolute; stem repeatedly trichotomously and diffusely branched. North of Kimball, August 12 [NEB 171171, 171172; RM 225453]. Not in Webber's CATALOGUE.

337. ***Eriogonum corymbosum*** Benth. [*E. nebraskense* Rydb.] – It is not given in Webber's CATALOGUE, and is not found in the Herbarium [NEB] of the University. Prairie: Kimball, August 12 [NEB 171172 **isotype** of *Eriogonum nebraskense* Rydb.; NY 323405 **holotype**, 323406 **isotype**; US 63360 **isotype**]; Potter, Cheyenne Co., July 15*.

J. M. Holzinger: "*Eriogonum microthecum*, a more densely flowered form."

*The July 15 date is incorrect because Rydberg and Flodman were in northern Banner Co. that day, about 60 miles from Potter. The *Report* notes this plant at Potter on August 14, and perhaps that was the intended date.

338. ***Eriogonum brevicaule*** Nutt. – Collected for the first time in Nebraska, this year. Hills in Kiowa Valley, Scotts Bluff Co., July 28 [NEB 171137] & 29. [NEB 207860 is August 1, at this site.]

339. ***Rumex venosus*** Pursh – Hills and sandy soil: Near Chappell, Deuel Co., June 29 [NEB 171852]; Court House Rock, Cheyenne Co., July 4.

340. ***Rumex salicifolius*** Weinman [*R. triangulivalvis* (Danser) Rech. *f.*] – Sands of [North] Platte [River], near Horse Creek, Scotts Bluff Co., August 1 [NEB 171834].

341. ***Rumex*** sp. . . . . . *sic* – I do not know where to place this. It is not *R. altissimum*, as all the valves are grain-bearing. The general appearance is too unlike *R. britannica*. The valves are not subrhomboid as in *R.*

verticillatus, and the pedicels are shorter. The size of the plant and the form of the panicle and of the valves are those of *R. salicifolius*, but the leaves reticulate and somewhat wavy-margined in the manner of *R. maritimus*. It seems to be intermediate between the preceding and the following, together with which two it grew on sands of the Platte River, near Horse Creek, Scotts Bluff Co., August 1. [Specimen not at NEB, but see 340.]

J. M. Holzinger: "*Rumex crispus*".

342. ***Rumex maritimus*** L. [*R. fueginus* Philippi] – Sands of [North] Platte [River], near Horse Creek, Scotts Bluff Co., August 1 [NEB 171781].

343. ***Polygonum erectum*** L. [*P. buxiforme* J. K. Small] – Unusally large and decumbent specimens, growing in water in Lodgepole Creek, near Sidney, Cheyenne Co., August 18 [NEB 171429].

344. ***Polygonum*** sp. . . . . *sic* [*P. aviculare* L.] – Stem much branched, low, prostrate, branchlets upright with very short joints, nearly covered by the sheaths, which are brown at the base; the scarious tips long and loose, soon lacerate, leaves crowded, oblong, 2–3 lines long, deciduous; flowers minute, reddish or brownish; seeds shining. It agrees with the description of *P. minimum* Wats. or still better to that of *P. torreyi* Wats., but neither is reported from as far east as Nebraska. It, maybe, is but a small variety of *P. aviculare*. The locality indicates, however, that it without any doubt [is] indigenous. Near a spring, in Deuel C., August 24 [NEB 171240, 171429].

J. M. Holzinger: "A remarkable little *Polygonum*, collected by Wright in 1851 in N. Mex., and by Newberry (?) *sic* near Dallas, Texas. I make it *Polygonum aviculare* var. *parva, var. nov.* Its small size, short internodes and conspicuous, strongly ribbed, and aristate margined brown ochreae give it a very characteristic appearance. Yet I cannot make it a species distinct from *aviculare*."

345. ***Polygonum ramosissimum*** Michx. – Very variable. Sand-draw, Deuel Co., August 24 [NEB 171262]; Lawrence Fork, Banner Co., July 8; Carter Canyon, Scotts Bluff Co., July 26 [NEB 171516].

346. ***Polygonum tenue*** Michx. [probably *P. sawatchense* J. K. Small] – Hills in Banner Co., July 10.

347. ***Polygonum lapat[h]ifolium*** L. – Lodgepole Creek, near Kimball, August 12 [NEB 171403].

348. ***Polygonum lapat[h]ifolium*** L. var. ***incanum*** Koch – Lawrence Fork, Banner Co., July 8.

349. ***Polygonum muhlenbergii*** Wats. [*P. coccineum* Muhl. ex Willd.] – [North] Platte River, near Horse Creek, Scotts Bluff Co., August 1 [NEB 171288]; Lodgepole Creek, near Kimball, August 12.

350. ***Polygonum convolvulus*** L. – Railroad bank, near Kimball, Aug. 12 [NEB 171276]

351. ***Polygonum dumetorum*** L. var. ***scandens*** Gray [*P. scandens* L.] – The wings of the fruits of nearly all my specimens are more or less lobed. McCuligan Canyon, Deuel [now Garden] Co., August 26* [NEB 171628, 171633].

*The *Report* cites August 22 as the day of collection. On August 26, Rydberg was in Chappell, about 20 miles from McCuligan Canyon.

[ELAEAGNACEAE]

352. ***Shepherdia argentea*** Nutt. – In fruit, on the islands in the Platte River, near Gering, Scotts Bluff Co., July 24* [NEB 180563].

*The *Report* gives July 21 as the date, but the label on the specimen clearly shows July 24. Nevertheless, July 21 is likely correct because

Rydberg and Flodman were on the North Platte River bottoms then, and on July 24 they were in nearby badlands.

EUPHORBIACEAE

353. ***Euphorbia petaloidea*** Eng. [*E. missurica* Raf. var. *petaloidea* (Engelm.) L. C. Dorn] – Variable; leaves from linear to broadly oblong; appendages from oval oblong to nearly triangular and small; seeds from gray to reddish. Sand-draw, Deuel Co., June 25 & Aug. 4; Scotts Bluff, July 22; Kiowa Valley, July 30; McCuligan Canyon, Deuel [now Garden] Co., August 22. [NEB 207859 is *Rydberg s.n.*, Lawrence Fork, Banner Co., July 10, 1891].

Rydberg's *Report* mentions this plant for several other dates from these and other sites. He also collected it in Banner Co. in August, 1890 (NEB 179037).

354. ***Euphorbia serpyllifolia*** Pers. – Railroad bank, Curtis, Frontier Co., June 22; Kiowa Valley, Scotts Bluff Co., July 30.

355. ***Euphorbia glyptosperma*** Eng. – Lawrence Fork, Banner Co., July 8; Kiowa Valley, Scotts Bluff Co., July 28; sands of Platte, near Horse Creek, August 1; McCuligan Canyon, Deuel [now Garden] Co., August 26* [NEB 178700]; sands of South Platte, Big Springs, August 25.

*The *Report* cites August 22 for the McCuligan Canyon specimen, but the label on the specimen clearly reads August 26; the former date is correct because Rydberg was in McCuligan Canyon that day and in Chappell, 20 miles away, on August 26. The *Report* does not cite the July 8 and July 28 specimens.

356. ***Euphorbia*** sp. . . . . *sic* [*E. serpyllifolia* Pers.] – It is near to *P. serpyllifolia* & *P. glyptosperma*; upright, much branched; leaves linear oblong a little oblique at the cuneate base, short petioled, serrate at the apex, about ½ inch long; peduncles axillary, leafy cluster; pods angled, seeds oblong, brownish, 4-angled, minutely wrinkled or pitted. The general appearance is somewhat like that of *P. glyptosperma*, from which it differs in the seeds. Cultivated soil, 6 miles south of Potter, Cheyenne Co., August 13. [NY 263066, **holotype** of *Euphorbia albicaulis* Rydb.; US 65781, **isotype** of *Euphorbia albicaulis* Rydb.].

J. M. Holzinger: "I read your note with care, examined our plants and compared with Herbarium specimens before deciding. Yet, I may have erred. And Dr. Millspaugh, who is the specialist on *Euphorbias*, may correct the determination. I have supposed your plant to be a western variety of *serpyllifolia*."

This collection is not mentioned in Rydberg's *Report*.

357. ***Euphorbia maculata*** L. – The leaves are unusually broad as in *E. humistrata*, and the involucre silky; the seeds oblong, sharply angled, with about 4 blunt ridges. Upper Lawrence Fork [Kimball Co.], August 12.

358. ***Euphorbia fendleri*** T. & G.– Upper Lawrence Fork, Kimball Co., August 12 [NEB 178670, collected by Flodman]. Collected for the first time in Nebraska, by me in 1890 [Cheyenne Co., NEB 178671, 178672].

359. ***Euphorbia hexagona*** Nutt. – On sand, common. Lawrence Fork, Banner Co., July 8; McCuligan Canyon, Deuel [now Garden] Co., August 22 [26!]; sand-draw, Deuel Co., August 23 [NEB 178763].

Rydberg also collected this species in Banner Co. in August, 1890 (NEB 178766).

360. ***Euphorbia marginata*** Pursh – Near Sidney, Cheyenne Co., August 18.

361. ***Euphorbia cuphusperma*** Boiss. (?) *sic* [*E. davidii* Subils] – All specimens I collected last year were smooth, wherefore I did not believe that they belonged to *E. dentata*. As the seeds are not ovate-globular, but ovate pyramidal, more or less 3–4-angled, obliquely truncate at the base, grooved on the shorter side, and rather prominently

tuberculate, so I named it *E. cuphusperma*, although this has only been collected at the Mexican Boundary. All specimens collected this year, at the same station, are more or less hairy, and the leaves broader and more dentate. It is undoubtedly nearly related to *E. dentata*, if not a variety of it. The leaves are linear or linear lanceolate, seldom broader, slightly dentate, and somewhat revolute.–Canyon, near Lawrence Fork, Banner Co., July 9 [US 91611].

J. M. Holzinger: "Your determination is correct; it is *Euphorbia cuphusperma* Boiss. The seeds are exactly like ours from Wright's Mexican plants, coll. in 1834."

Rydberg also collected this species in Banner Co. in August, 1890 (NEB 178603, US 91611).

362. ***Euphorbia montana*** Eng. [*E. robusta* Small ex Britt.] – Hills: Near McCuligan Canyon, Deuel [now Garden] Co., June 26; near Lawrence Fork, Banner Co., July 8 [NEB 178840].

Rydberg also collected this species in Banner Co., August, 1890 (NEB 207930).

363. ***Croton texensis*** Müll. – Sand: Lawrence Fork, Banner Co., July 8 [NEB 178528]; Rush Creek, Deuel [now Garden] Co., July 2 [US 102258] ; Potter, Cheyenne Co., August 16.

Rydberg also collected this species in sand draws of Deuel Co. in July, 1890 (NEB 178578).

URTICACEAE

364. ***Celtis occidentalis*** L. – Not uncommon in canyons and near water. In some rocky places it becomes only a small tree, with narrow leaves and smooth, not corky bark. Sidney, Cheyenne Co., August 18.

Rydberg mentioned this plant numerous times in his *Report*, sometimes as *C. canadensis*. It is common and the only species of *Celtis* in the area. He also collected it at Hackberry Springs, Banner Co., in August, 1890 (NEB 170749).

365. ***Urtica gracilis*** Ait. [*U. dioica* subsp. *gracilis*] – Common throughout the state.– Canyon, near Big Horn, Banner Co., August 8.

366. ***Parietaria pensylvanica*** Muhlenberg – Shady places: Curtis, Frontier Co., June 22; Lawrence Fork, Banner Co., July 9 [RM 248313].

PLATANACEAE

367. ***Platanus occidentalis*** – Some young trees (maybe raised from seeds) on the bank of Platte River in Kearney Co. It has been reported from only the very S.E. corner of Nebraska. July 15* [June 18] [NEB 174650]

* While the label notes July 15, that date is incorrect because Rydberg and Flodman were in Banner Co. that day, about 200 miles from Kearney Co. The *Report* discusses this plant on June 18, when they were in Kearney and Phelps counties. These trees were several miles inside Phelps Co. and not in adjacent Kearney Co. The specimen is surely from a cultivated plant because, as Rydberg noted, in Nebraska the tree is native only in far southeastern counties more than 200 miles to the east.

SALICACEAE

368. ***Salix longifolia*** Muhl. [*S. interior* Rowlee, *S. exigua* Nutt. subsp. *interior* (Rowlee) Cronq.] – A very variable species, common. Lawrence Fork, July 9 [8] [NY 73754, **holotype** of *Salix luteosericea* Rydb.]; on [North] Platte bottoms, near Horse Creek [Scotts Bluff Co.], August 1 [NEB 170257, RM 2580]; and in Mitchell Valley, Scotts Bluff Co., August 4; Kimball, August 23; Sidney, at Lodgepole Creek, August 18.

J. M. Holzinger: "368, 369, 370. About the willows you may hear again, since I wish to refer all the Salices to Mr. Bebb."

The *Report* mentions this plant numerous times and from many places.

369. ***Salix longifolia*** Muhl. var. . . . . *sic* [*S. interior* Rowlee] – Leaves shorter; female

aments short, capsule short, ovate. Only a small bush, very often monoecious. Platte [River], Kearney Co., June 15; canyon, Gosper Co., June 20; Carter Canyon, Scotts Bluff Co., July 26; Horse Creek, August 1.

370. ***Salix longifolia*** Muhl. var. . . . . *sic* – Stem blue. Carter Canyon. [See 369 above]

371. ***Salix cordata*** Muhl. var. ***vestita*** And. [*S. amygdaloides* Michx.] – Lawrence Fork, Banner Co.

372. ***Populus*** sp. . . . . . *sic* [*P.* × *acuminata* Rydb.] – When approaching the Carter Canyon in Scotts Bluff County, I saw at a distance some dark green trees with pyramidal crown. Not knowing any other tree with dark foliage and pointed top growing in the region, I took it for unusually tall specimens of *Juniperus virginiana*. Coming a little nearer I saw my mistake. It was a *Populus* unlike all I have seen. As the buds were very balsamiferous and the leaves cuneate at the base, I thought it was nothing but *P. angustifolia*, which I have never seen growing. When at home, I compared it with my specimens of this poplar, collected by Mr. T[homas] A. Williams in the Hat Creek Basin in northwestern Nebraska, and I saw at once the difference between the two. Although the leaves of my poplar are rhomboid ovate, cuneate at the base, and not whitened beneath, do I (*sic*) believe that, if not distinct from both, it is a variety of *P. balsamifera* rather than of *P. angustifolia*. The leaves are rhomboid, on petioles 1–2 inches long (in *P. angustifolia*, they are only ½–1 inch), generally long acuminate, shining on both sides, regularly crenate except at the cuneate base. The teeth are larger, more distant, and more regular than in *P. angustifolia*. Dr. Bessey states that the general appearance of the true *P. angustifolia* is more that of a willow than that of a poplar, which is not at all the case with this. The general growth, the size, form, and color of the trunk were somewhat between those of *P. balsamifera* and *P. angulata [angustifolia]*, although most like the former. The long and slender petioles reminded of the latter or even of quaking aspen. Dr. S. Watson says in his Revision about *P. angustifolia*: "Two forms are spoken of, the yellow cottonwood, making fair lumber, and the black cottonwood, common and extensively planted in Utah, but the wood is considered worthless." Maybe this is the former. I saw it growing only at one place, viz., in Carter Canyon, where it grew along the brook, together with *P. angulata* [*angustifolia*]. The grove contained over 100 trees of this poplar, from 40 to 60 feet high, some measuring 18 inches in diameter.– Carter Canyon, Scotts Bluff Co., July 26* [GH 30820 **isotype** of *Populus acuminata* Rydb.; NY 283994 **type** of *Populus acuminata* Rydb., labeled July 25; US 15164 **isoytpe,** labeled July 26]. [Rydberg's specimens are not at NEB, but specimens from other collectors have since been collected in Carter Canyon].

J. M. Holzinger: "I make after careful consideration *Populus angustifolia* James. The difficulty seems to arise from the mature leaves: herb. material generally has them undeveloped."

*These trees are discussed in Rydberg's *Report* for July 25, and it is unlikely that the specimens were collected on July 26 because that was a Sunday in 1891, and Rydberg and Flodman did not collect on Sundays.

IRIDACEAE

373. ***Sisyrinchium aceps*** Cav. [*S. montanum* Greene] – Near a spring, south of Pumpkin Seed Valley, Banner Co., July 6 [NEB 169137].

The July 6 collection is not mentioned in the *Report*, but the species was collected at that location on July 13. It is the only species of *Sisyrinchium* in range. Rydberg and Flodman visited the site on both days.

LILIACEAE

374. ***Allium reticulatum*** Fraser [*A. drummondii* Regel] – It is not recorded in

Webber's CATALOGUE. I found it growing at several places on the prairies; it was out of bloom and only a few specimens were collected.–Prairie in Deuel Co., June 25.

John M. Holzinger: "*Allium nuttallii*"

375. ***Yucca angustifolia*** Pursh [*Y. glauca* Nutt. is the only native species in Nebraska.] – On the hills of western Nebraska. Canyon in Gosper Co., June 20; Scotts Bluff, July 25 [NEB 169520].

Rydberg also collected this species in Banner Co. in August, 1890 (NEB 169517). The plant is illustrated in the frontispiece and on the cover, growing among ponderosa pines in the Wildcat Hills.

376. ***Fritellaria atro-purpurea*** Nutt. *sic* [*Fritillaria atropurpurea*] – In fruit only, near Scotts Bluff, July 23.

NEB 169538 is from Banner Co. on July 6 and is mounted on same sheet as *Zigadenus venenosus*, no. 377 below.

377. ***Zygadenus*** sp. . . . . . *sic* [*Zigadenus venenosus* S. Watson var. *gramineus* (Rydb.) Walsh ex M. E. Peck; *Toxicoscordion gramineum* (Rydb.) Rydb.] – In fruit only. As I have not seen it in bloom, and as there are no specimens in the Herbarium [NEB] like it, I cannot determine it but believe that it belongs to this genus. Hills of Pumpkin Seed Valley, Banner Co., July 6 [NEB 169538].

John M. Holzinger: "*Zygadenus nuttallii* Gray."

This specimen, the only *Zigadenus* in range, is mounted on the same sheet as *Fritillaria atropurpurea*, no. 376 above.

JUNCACEAE

378. ***Juncus filiformis*** L. (?) *sic* [probably *J. arcticus* Willd. var. *balticus* (Willd.) Trautv.] – North Platte River, Cheyenne [now Morrill] Co., July 3.

John M. Holzinger: "*Juncus balticus*"

379. ***Juncus tenuis*** Willd.–Wet prairies, Kearney Co., June 15. – Pumpkin Seed Valley, Banner Co., July 16. 379b is a variety with channeled leaves, lax panicle, and very long bracts. Pumpkin Seed Valley, Banner Co., July 6.

Rydberg also collected this species near Lawrence Fork, Banner Co. in 1890 (NEB 169153, 169219).

380. ***Juncus bufonius*** L. – It is not given in Webber's CATALOGUE, but has been collected in Nebraska before, and by me in 1890*. North Platte River, Cheyenne [now Morrill] Co., July 3 [NEB 168994]; Carter Canyon, Scotts Bluff Co., July 26 [NEB 155528, 168982, 168992]; Canyon near Big Horn, Banner Co., August 8.

The *Report* also notes this plant for Pumpkin Seed Valley, Banner Co., on July 14, 1891. *Lawrence Fork, Banner Co., August, 1890: NEB 168981 and 168995, identified by Frederick V. Coville in 1894.

381. ***Juncus nodosus*** L. var. ***megacephalus*** Torr. – Platte River near Horse Creek, Scotts Bluff Co., August 1.

Rydberg also collected this species at Lawrence Fork, Banner Co. in 1890 (NEB 169184).

382. ***Juncus nodosus*** L. var. . . . . . . (?) *sic* – Sidney, Cheyenne Co., Aug. 13.

John M. Holzinger: "*Juncus nodosus* var. *megacephalus*, = 381"

TYPHACEAE

383. ***Typha latifolia*** L. – In Lodgepole Creek, east of Potter, Cheyenne Co., August 17.

384. ***Sparganium eurycarpum*** Eng. – The leaves before drying were nearly triangular. In Lodgepole Creek, east of Potter, Cheyenne Co., August 17 [NEB 168306, 168307].

LEMNACEAE

385. ***Lemna minor*** L. – Lawrence Fork, Banner Co., July 10 *sic* [July 11 on label] [NEB 248587].

386. ***Lemna perpusilla*** Torr. (?) *sic* [*L. minor* L.] – It might be a variety of *L. minor* due to the condition of the water in which it grew. The frond is oblong, with distinct although not prominent nerves. The root is slender, up to 2 inches long. Spring near Horse Creek, Scotts Bluff Co., August 1 [NEB 248588].

John M. Holzinger: "*Lemna minor.*"

ALISMACEAE

387. ***Alisma plantago*** L. var. ***americana*** Gray [*A. triviale* Pursh] – Low ground, Mitchell Valley, Scotts Bluff Co., August 4 [NEB 168684, 168685].

388. ***Sagittaria variabilis*** Eng. [*S. cuneata* E. Sheld.] – Pumpkin Seed Creek, Banner Co., July 16; Horse Creek, Scotts Bluff Co., August 1 [NEB 168775].

Rydberg also collected this plant in Lodgepole Creek, Deuel Co., in July, 1890 (NEB 55289).

NAJADACEAE

389. ***Zannichellia palustris*** L. – Lawrence Fork, in the creek, Banner Co., July 11 [NEB 168618, 168619].

The *Report* also mentions this plant for Pumpkin Seed Creek, Banner Co., on July 16, 1891.

390. ***Triglochin maritima*** L. – Low ground: Platte River, Kearney Co., June 15 [NEB 168627, 168641]; Pumpkin Seed Valley, Banner Co., July 17.

CYPERACEAE

391. ***Cyperus aristatus*** Rottb. [*C. squarrosus* L.] – South of Platte River near Horse Creek, Scotts Bluff Co., August 1 [NEB 192906, 193053].

392. ***Cyperus schweinitzii*** Torr. – Wet prairies: Kearney Co., June 15 [NEB 193098, 193103]; Deuel Co., June 25.

393. ***Scirpus pungens*** Vahl [*Schoenoplectus pungens* (Vahl) Palla] – The culm has often only one spikelet. Lawrence Fork, Banner Co., July 8 [NEB 193559, 193560].

Rydberg also collected this species at Hackberry Springs, Banner Co., in August, 1890 (NEB 193529).

394. ***Scirpus lacustris*** L. [*Schoenoplectus tabernaemontani* (C. C. Gmel.) Palla; Scirpus validus Vahl] – Canyon, Kiowa Valley, Scotts Bluff Co., July 30 [NEB 193720, 193725].

The *Report* also cites this plant for the North Platte River at Big Springs, Deuel Co., on August 25.

395. ***Scirpus maritimus*** L. [*Bolboschoenus fluviatilis* (Torr.) Soják] – Platte River, Kearney Co., June 15 [NEB 193607].

396. ***Scirpus atrovirens*** Muhl. var. ***pallidus*** Britton [*S. pallidus* (Britt.) Fernald] – Low ground, Kiowa Valley, Scotts Bluff Co., July 30 [NEB 193660].

The *Report* cites this plant for the North Platte River at Big Springs, Deuel Co., on August 20.

397. ***Eliocharis*** *sic* ***acicularis*** R. Br. – Wet prairie, Deuel Co., June 27 [NEB 193251]; Lodgepole Creek, Cheyenne Co., August 18 [NEB 193231, 193258].

398. ***Eliocharis*** *sic* sp. . . . .[*Eleocharis compressa* Sullivant] – Culm somewhat flattened, striate; achene obovate, muricate. It might be a variety of *E. compressus*, which has not been collected in Nebraska, but the achenes are not strictly triangular, although a little thickened on the back. Dr. Britton mentions in his Revision that the specimens of *E. palustris* with flattened

culm have been collected in Mo.–Texas. This is not *E. palustris*, at least not *E. palustris* of Europe. Wet prairie, Kearney Co., June 13 [NEB 193258, 193269].

John M. Holzinger: "*Eliocharis (sic) palustris.*"

399. ***Fimbristylis spadicea*** Vahl. Var. ***castanea*** Gray [*F. puberula* (Michx.) Vahl] – It is not given in Webber's CATALOGUE, but was collected in 1889 by Rev. Bates of Valentine, Neb. Scotts Bluff, July 23 [NEB 193475]; near Horse Creek, Scotts Bluff Co., August 1 [NEB 193473].

400. ***Carex*** sp. . . . . . . [*C. filifolia* Nutt.] – This is named "bunch grass," "wire grass," etc. The first name it received because it grows in very dense, hard tufts, the last because its fine roots are black and, when dry, tough as wire. The ground on which it grows is regarded as the best, but it is very hard to break on account of the tufts, which do not rot for several years. It was out of bloom when I reached western Nebraska. Only one specimen was secured. I have not determined to which species it belongs. This is also the case with all my collections of the genus *Carex*. I have not classified a single American *Carex*. Supposing them as hard as the European ones, I leave them undetermined. Prairie, Deuel Co., June 25 [NEB 192032*]

John M. Holzinger: "*Carex filifolia.*"

*NY 2275873 is possibly a duplicate but has insufficient information beyond the printed label: Flora of Western Nebraska, Coll. P. A. Rydberg, 1891. The script (*Carex filifolia*, 1078) is not in Rydberg's hand, and the number, 1078, does not conform to his numbering on that trip. The specimen was perhaps part of Herbert Webber's private herbarium.

This sedge and its importance are discussed in the *Report* for June 13, June 27, August 10.

401. ***Carex*** . . . . . *sic* [*C. heliophila* Mack.] – Prairie, Deuel Co., June 27 [NEB 192492].

John M. Holzinger: "*Carex marcida* Booth. var. *debilis*"

402. ***Carex*** . . . . . *sic* [*C. brevior* (C. Dewey) Mack. ex Lunell] – Wet prairie, Kearney Co., June 13 [NEB 192081].

John M. Holzinger: "*Carex muhlenbergii.*"

403. ***Carex*** . . . . . *sic* [*C. gravida* L. H. Bailey] – Wet prairie, Kearney Co., June 13 [NEB 192048, 192051, 192060, 192064].

John M. Holzinger: "*Carex gravida.*"

404. ***Carex*** . . . . . *sic* [*C. nebrascensis* C. Dewey] – Lawrence Fork, Banner Co., July 10 [NEB 192450].

John M. Holzinger: "*Carex filiformis.*"

GRAMINEAE

405. ***Panicum capillare*** L. – Hill, Kiowa Valley, Scotts Bluff Co., July 30; Sidney, Cheyenne Co., August 18 [NY 1766911].

406. ***Panicum virgatum*** L. – Prairie, near Horse Creek, Scotts Bluff Co., August 1.

John M. Holzinger: "*Panicum virgatum* var. *glaucum*"

407. ***Panicum virgatum*** L. var. ***confertum*** Vasey [*Panicum virgatum* L.] – Not reported for Nebraska. Lawrence Fork, Banner Co., July 8.

408. ***Panicum scoparium*** Lam. [perhaps *Panicum leibergii* (Vasey) Scribn.] – Sandhills, Kearney Co., June 15.

409. ***Panicum dichotomum*** L. var. ***barbulatum*** (Michx.) Gray [perhaps *Panicum acuminatum* Sw.] – Not reported for Nebraska. Platte River, Kearney Co., June 15.

410. ***Panicum crus-galli*** L. var. . . . .[probably *Echinochloa muricata* (P. Beauv.) Fernald var. *micostachya* Wiegand] – In my belief, this is

indigenous, as it was found in a canyon several miles from any habitation. Furthermore, the country in Kiowa Valley has been very *little* under cultivation. In my specimens, the awns are short and the glumes hispid. Canyon in Kiowa Valley, Scotts Bluff Co., July 28.

John M. Holzinger: "*Panicum crus-galli* var. *muticus*."

411. ***Cenchrus tribuloides*** L. [*C. longispinus* (Hack.) Fernald] – Rare in the west, cultivated ground south of Potter, Cheyenne Co., August 13.

412. ***Spartina cynocuroides*** *sic* (L.) Willd. [*S. pectinata* Link] – Near water, Platte River, Scotts Bluff Co., July 23; Horse Creek, August 1.

Spartina gracilis was seen but not collected near Kimball, in Kimball County.

413. ***Leersia oryzoides*** Swartz. – Lodgepole Creek, east of Potter, Cheyenne Co., August 18 [NEB 196147].

414. ***Andropogon furcatus*** Muhl. [*A. gerardii* Vitman subsp. *gerardii*]–Common throughout the state. – East of Potter, Cheyenne Co., August 18 [NEB 194217].

This grass is mentioned in the *Report* for several places on different dates.

415. ***Andropogon scoparius*** Michx. var. . . . . . [*Schizachyrium scoparium* (Michx.) Nash] – This species is described in the last edition of Gray's MANUAL as having lower sheaths and narrow leaves hairy. All Nebraska specimens I have seen are wholly glabrous and somewhat glaucous. Lawrence Fork, Banner Co., July 10; Cedar Canyon, Scotts Bluff Co., July 22; east of Potter Cheyenne Co., August 18.

John M. Holzinger: "415, 431, 457: Dr. Vasey will write to you himself about these at his earliest convenience."

416. ***Andropogon hallii*** Hackel [*Andropogon gerardii* Vitm. subsp. *hallii* (Hack.) J. Wipff] – Hills of western Nebraska: Lawrence Fork, Banner Co., July 10 [NEB 194257]; Cedar Canyon, Scotts Bluff Co., July 22; east of Potter, Cheyenne Co., August 18.

This plant is mentioned in the *Report* for several other places on different dates.

417. ***Chrysopogon nutans*** Benth. [*Sorghastrum nutans* (L.) Nash] – Cedar Canyon, Scotts Bluff Co., July 22; Big Springs, Deuel Co., August 26 [NEB 197760].

418. ***Phalaris canariensis*** L. – Minden, Kearney Co., June 13 [NEB 197191, 197192].

419. ***Phalaris arundinacea*** L. – Platte River, Kearney Co., June 15 [NEB 197171].

420. ***Alopecurus geniculatus*** L. [*A. carolinianus* Walter] – Wet prairie, Deuel Co., June 27 [NEB 194173].

421. ***Alopecurus aristulatus*** Michx. [*A. aequalis* Sobol.] – Carter Canyon, Scotts Bluff Co., July 26 [NEB 194145].

422. ***Aristida purpurea*** Nutt. – Hills: Curtis, Frontier Co., June 22; Kiowa Valley, Scotts Bluff Co., July 26; Sidney, August 19.

423. ***Aristida purpurea*** Nutt. var. ***hookerii*** Trin. [probably var. *purpurea*]– My specimens are like those of this variety in the Herbarium [NEB] of the University of Nebraska, but I cannot find any description.– Sandhills of Kearney Co., June 16.

424. ***Stipa comata*** Trin. & Rup. – It takes the place of *S. spartina* [*sic*] in the west.–Hills of Kearney Co., June 15 [NEB 198113]; prairies: Perkins Co., June 23, Deuel Co., June 26, Banner Co., July 12.

425. ***Stipa viridula*** Trin. [*Nasella viridula* (Trin.) Barkworth] – Sandhills of Phelps Co., June 18; McCuligan Canyon, Deuel [now

Garden] Co., June 26; North Platte [River], Cheyenne [now Morrill] Co., July 3 [NEB 198254].

426. ***Oryzopsis micrantha*** Thurber – A very delicate grass on rich soil. Lawrence Fork, Banner Co., July 8 [RM 224522] .

427. ***Oryzopsis cuspidata*** Benth.[*Stipa hymenoides* Roem. & Schult.] – The most beautiful grass in Nebraska. Not uncommon west.–McCuligan Canyon, Deuel [now Garden] Co., June 26 [RM 224521]; Rush Creek, Deuel [now Garden] Co., July 2; [North] Platte River, Cheyenne Co., July 3.

428. ***Muhlenbergia ramosa*** [*racemosa*] B.S.P. – Lodgepole, near Sidney [Cheyenne Co.], Aug. 19 [NEB 196540].

> John M. Holzinger: "*Muhlenbergia glomerata.*"

429. ***Muhlenbergia pungens*** Thurb. – It is called "sand grass" in Kearney Co.–Sandhills of Kearney Co., June 15; in the "bad lands" north of Scotts Bluff, July 24.

430. ***Sporobolus cuspidatus*** Torr. [*Muhlenbergia cuspidata* (Torr.) Rydb.] – Hills: Big Horn, Banner Co., August 8. [See no. 431.]

431. ***Sporobolus cuspidatus*** Torr. var. . . . . [*Muhlenbergia cuspidata* (Torr.) Rydb.] – This is undoubtedly nothing but a low tufted form of *S. cuspidatus*, although the lower glumes are only half the length of the upper. Hills: Freeport, Banner Co., July 15; Big Horn, August 8 [NEB 196396].

> John M. Holzinger: "Dr. Vasey will write to you himself about these at his earliest convenience."

432. ***Sporobolus cryptandrus*** (Trin.) Gr. – Platte River, Kearney Co., June 14; McCuligan Canyon, Deuel [now Garden] Co., June 26; Potter, Cheyenne Co., August 18.

433. ***Sporobolus cryptandrus*** var. . . . . *sic* – Culm strict, slender; leaves more or less involute; panicle long and open, branches mostly alternate and capillary, naked below. Dry sandy soil near Scotts Bluff, July 24 [NEB 197951; RM 222191].

> John M. Holzinger: "*Sporobolus cryptandrus* var. near *flexuosus.*"

434. ***Sporobolus airoides*** Torr. – Common on low ground. [North] Platte River, Cheyenne [now Morrill] Co., July 3 [RM 2602, 22172]; Pumpkin Seed Valley, Banner Co., July 16; Horse Creek, Scotts Bluff Co., August 1; Sidney [Cheyenne Co.], August 19.

435. ***Sporobolus asperifolius*** H. M. [*Muhlenbergia asperifolia* (Nees & Meyen ex Trin.) Parodi] – Carter Canyon, Scotts Bluff Co., July 26.

436. ***Sporobolus serotinus*** var. . . . . (?) *sic* [*Muhlenbergia minutissima* (Steud.) Swallen] – Depauperate specimens of a *Sporobolus*, undoubtedly of the named. Sands of [North] Platte River, near Horse Creek, Scotts Bluff Co., August 1 [NEB 312300].

> John M. Holzinger: "*Sporobolus confusus* Towen."

> This is the only specimen of the species ever collected in Nebraska, but it is known in adjacent Lararmie and other counties in Wyoming.

437. ***Agrostis alba*** L. [*A. stolonifera* L.] – The "red-top" in the East regarded as one of the best grasses for hay, is regarded by western stockmen nearly as good for nothing.–[North] Platte River, near Horse Creek, Scotts Bluff Co., August 1.

438. ***Calamagrostis longifolia*** Hook. [*Calamovilfa longifolia* (Hook.) Scrib.] – Low ground: Pumpkin Seed Valley, Banner Co., July 16 [NEB 195093]; Carter Canyon, July 26; near [North] Platte [River], July 23; near Horse Creek, August 1.

439. ***Calamagrostis canadensis*** Michx. – Platte River in Kearney Co., June 15 [NEB 194974, 194975].

440. ***Calamagrostis stricta*** Trin. – [North] Platte River, Cheyenne [now Morrill] Co., July 3 [NEB 195053]; Scotts Bluff Co., July 24.

441. ***Calamagrostis confinis*** Nutt. (?) [*C. stricta* (Timm.) Koeler] – Not reported for Nebraska. Platte River, Kearney Co., June 15 [NEB 194991, 195023].

442. ***Avena elatior*** [*Arrhenatherum elatius* P. Beauv. ex J. Presl & C. Presl] – New to Nebraska.–Lincoln [Lancaster Co.], June 12 [the day Rydberg left Lincoln to start the trip] [NEB 194510, 194514, probably from cultivation].

John M. Holzinger: "*Arrhenatherum avenaceum* Beauv."

443. ***Schedon[n]ardus texanus*** Steud. [*S. paniculatus* (Nutt.) Trel.] – [North] Platte River, Cheyenne [now Morrill] Co., July 3 [NEB 197581]; north of Kimball [Kimball Co.], August 12.

Rydberg spelled the generic name several ways in the *Report.*

444. ***Bouteloua oligostachya*** Torr. [*B. gracilis* (Willd. ex Kunth) Lag. ex Griffiths] – It is regarded as the best forage plant of the high plains and furnishes good pasturage even in the winter.–Rush Creek, Deuel [now Garden] Co., July 3.

445. ***Bouteloua hirsuta*** Lag. – Hills: Lawrence Fork, Banner Co., July 9; Upper Lawrence Fork, Kimball Co., August 12 [RM 2723].

446. ***Bouteloua racemosa*** Lag. [*B. curtipendula* (Michx.) Torr.] – Hills: Rush Creek, Banner [Garden!] Co., July 3* [*sic,* July 2] [RM 2725].

The part of Rush Creek where Rydberg collected is in Garden Co., and there is no Rush Creek in Banner County.

* The Report cites this collection for July 2, and on July 3 Rydberg was in adjacent Morrill Co.

447. ***Bouteloua dactyloides*** Eng. [*Buchloë dactyloides* (Nutt.) Engelm.] – The 'buffalo grass" is slowly disappearing from eastern Nebraska, but it is yet one of the principal grasses of high plains. Phelps Co., June 18 [NEB 194901], Kearney Co., June 13 [NEB 194904; NY 1643052].

448. ***Phragmites communis*** Trin. [*P. australis* (Cv.) Trin. ex Steud.] – Platte near Horse Creek, Scotts Bluff Co., August 1.

449. ***Munroa squarrosa*** Torr. – Frequents the "prairie dog towns."–Sand, Rush Creek, Deuel [now Garden] Co., July 2; Big Horn, Banner Co., August 8.

450. ***Koeleria cristata*** Pers. [*K. macrantha* (Ledeb.) Schult.] – Low specimens from the prairies of Kearney Co., June 13; Deuel Co., June 25 [NEB 196248].

451. ***Koeleria cristata*** Pers. var. ***gracilis*** Gray – Prairie, Deuel Co., June 25.

452. ***Eatonia obtusata*** Gray [*Sphenopholis obtusata* (Michx.) Scrib.] – Prairies, Wild Cat Mts., Banner Co., July 6; north of Scotts Bluff, July 23.

453. ***Eatonia*** sp. . . . . *sic* [*Sphenopholis obtusata* (Michx.) Scrib.] – The specimens are not fully developed. The leaves are broad (3-4 lines), ligules conspicuous, truncate, toothed and divided, the spike is more compound, with longer, more open branches; spikelets and the whole plant more robust than E. obtusata, which it otherwise resembles. Canyon in Kiowa Valley, Scotts Bluff Co., July 28 [NEB 195334].

John M. Holzinger: "*Eatonia obtusata* var. *robusta.*"

454. ***Catabrosa aquatica*** (L.) Beauv. – Lodgepole, near Sidney, August 19 [NEB 195108].

455. ***Distichlis spicata*** Green var. ***stricta*** Thurber [*D. spicata* (L.) Greene] – Salt marshes, railroad bank, Curtis, Frontier Co., June 22 [NEB 195261]; Platte River, Cheyenne Co., July 3; Big Horn, Banner Co., August 8.

456. ***Poa alpina*** L. [*P. pratensis* L.] – It is not given in Webber's CATALOGUE, but was collected in 1889 by T. A. Williams in War Bonnet Canyon [Sioux Co.]. McCuligan Canyon, Deuel [now Garden] Co., June 26 [NEB 197270].

457. ***Poa tenuifolia*** Nutt. var. . . . . .(?) *sic* [*P. arida* Vasey] – It agrees with the description of this species, but it is not densely tufted; the panicle is rather close; the rays are not distant, spreading. It is not uncommon on low land. Kearney Co., June 13 [NEB 197499]; Cheyenne Co., July 3.

John M. Holzinger: "Dr. Vasey will write to you himself about these at his earliest convenience."

458. ***Poa nevadensis*** (?) *sic* [*P. secunda* J. Presl] – This grows on drier soil and is more tufted than the preceding. Lawrence Fork, Banner Co., July 8 [NEB 197500].

John M. Holzinger: "*Poa tenuifolia* var. ?"

459. ***Glyceria distans*** (L.) Wahl. [*Puccinellia nuttalliana* (Schult.) A. S. Hitchc.] – Lodgepole, near Sidney, Aug. 19 [NEB 197519].

John M. Holzinger: "*Glyceria distans* var. *airoides*"

460. ***Glyceria nervata*** Trin. [*G. striata* (Lam.) A. S. Hitchc.] – Lodgepole, near Potter, Cheyenne Co., Aug. 18.

461. ***Glyceria*** sp. . . . . . *sic* [*Poa glaucifolia* Scribner & Williams] – Only a few undeveloped specimens were collected in Lawrence Fork, Banner Co., July 9. It seems to be somewhat near *G. obtusa*, but the panicle is smaller, more open, and the glumes more acute. (NY 431306 **holoype** of *Poa plattensis* Rydb., dated July 8 on NY website**)**

John M. Holzinger: "*Poa* sp. (insufficient)."

462. ***Festuca tenella*** Willd. [*Vulpia octoflora* (Walter) Rydb., *Festuca octoflora* Walter] – Sandhills, Kearney Co., June 13 [NEB 196068]; ...[site illegible] Deuel Co., June 25.

463. ***Agropyrum glaucum*** R. & S. var. . . . . *sic* [*Elymus smithii* (Rydb.) Gould, *Pascopyrum smithii* (Rydb.) Á. Löve] –The species of the genus *Agropyrum* are very difficult to distinguish. Especially is this the case with *repens*, *glaucum*, and *tenerum*, which all seem to run together. I have given each form a separate number; not that I regard them distinct, but mere forms.–In the first, the stem is slender, the spike narrow, the leaves involute, and the glumes pointed. It approaches somewhat *A. tenerum*. Scotts Bluff, July 24.

464. ***Agropyrum glaucum*** R. & S. var. . . . . *sic* [*Elymus lanceolatus* subsp. *lanceolatus*] – More stout, leaves involute, glumes blunt. Lawrence Fork, Banner Co., July 8 [RM 2705].

465. ***Agropyrum glaucum*** R. & S. var. . . . . *sic* – Stout, leaves involute, glumes pointed.–This is the most common form. Lawrence Fork, Banner Co., July 8; Prairie, Phelps Co., June 18.

466. ***Agropyrum glaucum*** R. & S. var. . . . . *sic* – Leaves flat, glumes acute or cuspidate. Low prairie, Kearney Co., June 13.

467. ***Agropyrum*** sp. . . . . *sic* – Probably a variety of *A. glaucum*, with the spikelets distorted by ergot. The leaves are short, involute, the spikelets few, many-flowered and crowded; glumes less plainly nerved, downy, acute. Kimball, August 12.

468. ***Agropyrum tenerum*** Vasey var. . . . . *sic* [*Elymus trachycaulus* (Link) Gould ex Shinners] – Weak and slender, leaves involute, glumes cuspidate or awned. McCuligan Canyon, Deuel Co., June 26; Pumpkin Seed Valley, Banner Co., July 6; Carter Canyon, Scotts Bluff Co., July 26.

469. ***Agropyrum tenerum*** Vasey var. . . . . *sic* [*Elytrigia dasystachya* (Hook.) Á. Löve & D. Löve; *Elymus lanceolatus* Scribn. & J. G.Smith) Gould] – Strict, leaves involute, setaceous, glumes blunt. Pumpkin Seed Valley, July 16. [US 1016374, **syntype** of *Agropyron divergens* var. *inerme* Scrib. & J. G. Smith]

470. ***Agropyrum tenerum*** Vasey var. . . . . *sic* – Slender, leaves broader, flat; glumes cuspidate or awned.–Lawrence Fork, Banner Co., July [6?, illegible]; canyon in Pumpkin Seed Valley, July 6; Carter Canyon, Scotts Bluff Co., July 26.

471. ***Agropyrum tenerum*** Vasey var. . . . . *sic* – It seems to be a luxurious form of *A. tenerum*, near the last, but with much larger spikelets, connecting it with the broad leaved forms of *A. glaucum*. Scotts Bluff, July 24.

John M. Holzinger: "*Agropyrum repens* var.?"

472. ***Elymus sitanion*** Schult. [*E. elymoides* (Raf.) Swezey ssp. *brevifolius* (J. G. Smith) Dorn, *Sitanion hystrix* (Nutt.) J. G. Smith] – It is not given in Webber's CATALOGUE, although it has been collected at several places in Nebraska. It seems to be not very uncommon, but is sometimes confounded with *Hordeum jubatum*. Prairie, Rush Creek, Deuel [now Garden] Co., July 2; Platte bottoms, Scotts Bluff Co., August 4 [NEB 197734].

473. ***Elymus canadensis*** L. – Very variable, sometimes with the spikes 6 in. long and the leaves ¾ inch wide. Low land near water: Lawrence Fork, Banner Co., July 8 [RM 2738]; [North] Platte River, Scotts Bluff Co., July 2; Mitchell Valley, Scotts Bluff Co., August 4.

CONIFERAE

474. ***Juniperus virginiana*** L. var. . . . . *sic* [*J. scopulorum* Sarg.] – The red cedar of eastern Nebraska is a good-sized tree, with conical crown, with the branches from near the ground. The leaves are as usually of two kinds, viz. on the younger branches awl-shaped, and on the older triangular in outline, but generally longer than broad. The red cedar of the lowlands and canyons of western Nebraska differs not materially from the eastern form, but on the Wild Cat Mountains and other crags in Banner and Scotts Bluff counties, there grew a low, scraggly tree, about 12–18 feet height, with rounded top, and the stem free from branches below. The color of the trunk was bright red, as in *Pinus resinosa*, and that of the branches grayish. The branchlets were 4-angled, crowded with very small, broadly triangular leaves, which are about as broad as long, resembling more those of *J. sabina* var. *procumbens*. I cannot remember that I saw any awl-shaped leaves at all. The berries are on straight peduncles as in *J. virginiana*, but smaller, scarcely 2 lines in diameter, with 2-angled and grooved seeds. Dr. Engelmann states that among the foothills of Pikes Peak, the trees of *J. virginiana* have the size and shape of an apple tree, with a rounded, spreading top. Dr. Bessey remarks that he also was struck by this peculiar form of the red cedars at Manitou, Col., which is not very far from Pike's Peak. This is undoubtedly the same variety. The red cedars of the North Platte region have also attracted the attention of Nebraska nurserymen, to which fact Dr. Bessey refers in his LIST OF TREES AND SHRUBS OF NEBRASKA (BESSEY, 1892).–Wild Cat Mountains, Banner Co., July 16 [NY 20817, as July 17].

John M. Holzinger: "Should also go with the species, in my judgment. The

Juniperus virginiana of Upper Wisconsin, along the St. Croix, has just such apple-tree like tops as you describe for your Nebraska form."

Rydberg saw but did not collect *Pinus ponderosa* var. *scopulorum*, which is abundant on the Wildcat Hills and Scotts Bluff. He mentioned it in his *Report* for June 26 and July 9, 14, 16, 24, and 28 as pine or Rocky Mountain yellow pine. It is the only pine at those sites.

EQUISETACEAE

475. ***Equisetum limosum*** L. var. .... *sic* [*E. laevigatum* A. Braun] – This has sometimes been named *E. laevigatum*, as the branches are very few and sometimes none. I took it at first for *E. robustum* or *hiemale*, but it lacks the black girdle.–Platte River, Kearney Co., June 15 [NEB 167934, 167998].

John M. Holzinger: "Should also go with the species, in my judgment."

476. ***Equisetum laevigatum*** Braun – Wet prairie, Kearney Co., June 13 *sic* [15 on the label] [NEB 167973, 167977].

477. ***Equisetum variegatum*** Schleicher [*E. laevigatum* A. Braun] – Wet prairies, Kearney Co., June 13 [NEB 168028].

FILICES

478. ***Aspidium thelypteris*** (L.) Sw. [*Thelypteris palustris* Schott var. *pubescens* (G. Laws.) Fernald] – Fronds only in a grassy marsh near Platte River, Kearney Co., June 15 [NEB 167785].

479. ***Cheilanthes lanuginosa*** Nutt. [*C. feei* Moore] – On exposed rocks, near Big Horn, Banner Co., August 8 [NEB 167680]. It is not given in Webber's CATALOGUE, but has been collected before in Nebraska, and by me in 1890 [Hackberry Springs, Banner Co., NEB 167678, 167679, 167681].

Appendix:

Summary of Nomenclature and Specimens

Rydberg's number	Name used by Rydberg	Modern or corrected name	Vouchers
309	*Abronia fragrans*	*Abronia fragrans*	NEB 172608
230	*Acerates auriculata*	*Asclepias engelmanniana*	NEB 182720, 182721
231	*Acerates viridifolia (sic)*	*Asclepias viridiflora*	NEB 182658
232	*Acerates viridifolia (sic)* var. *lanceolata*	*Asclepias viridiflora*	NEB 182633
202	*Achillea millefolium*	*Achillea millefolium*	NEB 199377
196	*Actinella acaulis*	*Tetraneuris acaulis*	NEB 190636
463	*Agropyron glaucum*	?	No specimen at NEB
464	*Agropyron glaucum* var. ___	*Elymus lanceolatus* subsp. *lanceolatus, Elytrigia dasystachya*	RM 2705
465	*Agropyron glaucum* var. ___	?	No specimen at NEB
466	*Agropyron glaucum var.* ___	?	No specimen at NEB
467	*Agropyron sp.*	?	No specimen at NEB
468	*Agropyron tenerum* var. ___	*Elymus trachycaulus*	No specimen at NEB
469	*Agropyron tenerum* var. ___	*Elymus lanceolatus* subsp. *lanceolatus, Elytrigia dasystachya*	US 1016374
470	*Agropyron tenerum* var. ___	?	No specimen at NEB
471	*Agropyron tenerum* var. ___	?	No specimen at NEB
437	*Agrostis alba*	*Agrostis stolonifera*	No specimen at NEB
387	*Alisma plantago* var. *americana*	*Alisma triviale*	NEB 168684, 168685
374	*Allium reticulatum*	*Allium drummondii*	No specimen at NEB
421	*Alopecurus aristulatus*	*Alopecurus aequalis*	NEB 194145
420	*Alopecurus geniculatus*	*Alopecurus carolinianus*	NEB 194173
103	*Amalanchier (sic) alnifolia*	*Amelanchier alnifolia*	NEB 175703
312	*Amarant(h)us blitoides*	*Amaranthus albus*	NEB 171889
313	*Amarant(h)us sp.*	*Amaranthus blitoides*	NEB 171934, 171937
178	*Ambrosia psilostachya*	*Ambrosia psilostachya*	NEB 187996
177	*Ambrosia trifida* var. *integrifolia*	*Ambrosia trifida* var. *integrifolia*	NEB 188030
63	*Amorpha canescens*	*Amorpha canescens*	NEB 176309
64	*Amorpha fruticosa*	*Amorpha fruticosa*	No specimen at NEB
39	*Ampelopsis quinquefolia*	*Parthenocissus vitacea*	NEB 179810, 179827
	Amphicarpea monoica	*Amphicarpaea bracteata*	In the *Report* but not the *List*
414	*Andropogon furcatus*	*Andropogon gerardii* subsp. *gerardii*	NEB 194217
416	*Andropogon hallii*	*Andropogon gerardii* subsp. *hallii*	NEB 194257
415	*Andropogon scoparius*	*Schizachyrium scoparium*	No specimen at NEB

Rydberg's number	Name used by Rydberg	Modern or corrected name	Vouchers
2	*Anemone cylindrica*	*Anemone cylindrica*	NEB 173267
173	*Antennaria plantaginifolia*	*Antennaria microphylla*	NEB 188257
288	*Aphyllon fasciculatum*	*Orobanche fasciculata*	No specimen at NEB
289	*Aphyllon fasciculatum* var. *luteum*	*Orobanche fasciculata*	NEB 186204
290	*Aphyllon ludovicianum*	*Orobanche ludoviciana*	NEB 186206
291	*Aphyllon ludovicianum*	*Orobanche ludoviciana*	No specimen at NEB
58	*Apios tuberosa*	*Apios americana*	In the *Report* but not the *List;* NEB 85701, June 15, '91, Kearney City
142	*Aplopappus armerioides*	*Stenotus armerioides*	NEB 191406, NY 260046
143	*Aplopappus nuttallii*	*Xanthisma grindelioides*	NEB 190781, 190783, 207815, 207820
145	*Aplopappus rubiginosus*	*Rayjacksonia annua*	NEB 128215, 128216, 128218, 207816
144	*Aplopappus spinulosus*	*Machaeranthera pinnatifida, Xanthisma spinulosum*	NEB 190820, 190822
226	*Apocynum cannabinum*	*Apocynum cannabinum*	NEB 182586
422	*Apocynum cannabinum*	*Apocynum cannabinum*	No specimen at NEB
12	*Arabis holboelii*	*Boechera grahamii*	NEB 174033
13	*Arabis* sp.	*Thelypodium integrifolium*	NEB 174644
31	*Arenaria franklinii var.*	*Arenaria hookeri, Eremogone hookeri*	NEB 172843
30	*Arenaria franklinii* var. *minor*	*Arenaria hookeri, Eremogone hookeri*	No specimen at NEB
32	*Arenaria* sp.	*Arenaria hookeri, Eremogone hookeri*	NEB 172838
9	*Argemone platyceras*	*Argemone polyanthemos*	NEB 173912
423	*Aristida purpurea* var. *hookeri*	*Aristida purpurea*	No specimen at NEB
205	*Artemisia biennis*	*Artemisia biennis*	No specimen at NEB
209	*Artemisia cana*	*Artemisia cana*	NEB 188359, 188360
203	*Artemisia canadensis*	*Artemisia campestris*	NEB 188448, 207827
204	*Artemisia dracunculoides*	*Artemisia dracunculus*	NEB 188426
206	*Artemisia filifolia*	*Artemisia filifolia*	NEB 188454
207	*Artemisia frigida*	*Artemisia frigida*	NEB 188482
208	*Artemisia ludoviciana*	*Artemisia ludoviciana*	NEB 188517, 188555, 207817
228	*Asclepias arenaria*	*Asclepias arenaria*	NEB 207845
227	*Asclepias speciosa*	*Asclepias speciosa*	NEB 192751
229	*Asclepias verticillata* var. *pumila*	*Asclepias pumila*	NEB 182859
478	*Aspidium thelypteris*	*Thelypteris palustris*	NEB 167785
162	*Aster adscendens (sic)*	*Aster ascendens*	No specimen at NEB
161	*Aster commutatus*	*Aster falcatus* susp. *commutatus*	NEB 207897
166	*Aster ericaefolius*	*Chaetopappa ericoides*	No specimen at NEB
165	*Aster foliaceus*	*Aster foliaceus*	NY 161978
163	*Aster fremontii*	?	No specimen at NEB
167	*Aster tanacetifolius*	*Machaeranthera tanacetifolia*	No specimen at NEB
164	*Aster tradescantii*	?	No specimen at NEB
69	*Astragalus adsurgens*	*Astragalus laxmanii*	NY 1203643, 1203645
78	*Astragalus caespitosus*	*Astragalus spatulatus*	NY 1283190
68	*Astragalus canadensis*	*Astragalus canadensis*	NY 1247095, 1247097
66	*Astragalus caryocarpus*	*Astragalus crassicarpus*	NEB 176747
71	*Astragalus gracilis*	*Astragalus gracilis*	NEB 176407; NY 1249873

Rydberg's number	Name used by Rydberg	Modern or corrected name	Vouchers
73	*Astragalus missouriensis*	*Astragalus missouriensis* var. *missouriensis*	NEB 176612; NY 1262031, 1262032
67	*Astragalus mollissimus*	*Astragalus mollissimus*	NEB 176692; NY 1265815; RM 243194
77	*Astragalus multiflorus*	*Astragalus multiflorus*	NEB 176969, 176704; NY 1266186; RM 223325
76	*Astragalus pectinatus*	*Astragalus pectinatus*	NEB 176664; NY 1246488
75	*Astragalus pictus* var. *filifolius*	*Astragalus ceramicus* var. *filifolius*	NEB 177916; NY 1247545
74	*Astragalus pubentissimus*	*Astragalus lotiflorus*	NEB 176636; NY 1261148, 1261149
70	*Astragalus racemosus*	*Astragalus racemosus* var. *racemosus*	NEB 176721; NY 1265591 1265604
81	*Astragalus sericoleucus*	*Astragalus sericoleucus*	NEB 178282, 1282679, 1282680
80	*Astragalus triphyllus*	*Astragalus hyalinus*	NEB 176846, US 69999
79	*Astraglus kentrophyta*	*Astragalus kentrophyta*	NEB 177011, NY 1258277
72	*Astraglus microlobus*	*Astragalus gracilis*	NEB 176421; NY 1249872, 1249875
325	*Atriplex argentea*	*Atriplex argentea*	NEB 172081, 172103
325 sic	*Atriplex nuttallii*	*Atriplex canescens*	NEB 172148, 172150; RM 234797
324	*Atriplex patula* var. *hastata*	*Atriplex dioica, Atriplex subspicata*	NEB 172113
56	*Aurea*	*Dalea aurea*	NEB 178013; NY 1073405, 1073406
442	*Avena elatior*	*Arrhenatherum elatius*	NEB 194510, 194514
198	*Bahia oppositifolia*	*Picradeniopsis oppositifolia*	NEB 190681
128	*Berula angustifolia*	*Berula erecta* var. *incisa*	NEB 181486
190	*Bidens cernua* var.	*Bidens cernua*	NEB 130283, 130294, 188738, 207806,
191	*Bidens chrysanthemoides* var. ___	*Bidens cernua*	NEB 188737
151	*Bigelovia graveolens* var. *glabrata*	*Ericameria nauseosa* var. *graveolens*	NEB 188869
150	*Bigelovia howardi (sic)*	*Ericameria parryi* var. *howardii*	NEB 188888; NY 214162
447	*Bouteloua dactyloides*	*Buchloë dactyloides*	NEB 194901, 194904; NY 1643031, 1643052
445	*Bouteloua hirsuta*	*Bouteloua hirsuta*	RM 2723
444	*Bouteloua oligostachya*	*Bouteloua gracilis*	No specimen at NEB
446	*Bouteloua racemosa*	*Bouteloua curtipendula*	RM 2725
137	*Brickelia* sic *grandiflora*	*Brickellia grandiflora*	NEB 188778, 188780, 188785, 207629, 207813; RM 221986
439	*Calamagrostis canadensis*	*Calamagrostis canadensis*	NEB 194974, 194975
441	*Calamagrostis confinis*	*Calamagrostis stricta*	NEB 194991, 195023
438	*Calamagrostis longifolia*	*Calamovilfa longifolia*	NEB 195093
440	*Calamagrostis stricta*	*Calamagrostis stricta*	NEB 195053
35	*Callirhoe alcaeoides*	*Callirhoe alcaeoides*	No specimen at NEB
34	*Callirhoe involucrata*	*Callirhoe involucrata*	NEB 179964
223	*Campanula rotundifolia*	*Campanula rotundifolia*	NEB 187017, 187033, 207842
401	*Carex*	*Carex heliophila*	NEB 192492
402	*Carex*	*Carex brevior*	NEB 192081
403	*Carex*	*Carex gravida*	NEB 192048, 192051, 192060, 192064
404	*Carex*	*Carex nebrascensis*	NEB 192450
400	*Carex* sp.	*Carex filifolia*	NEB 192032

Rydberg's number	Name used by Rydberg	Modern or corrected name	Vouchers
286	*Castileia (sic) sessiliflora*	*Castilleja sessiliflora*	NEB 185601
454	*Catabrosa aquatica*	*Catabrosa aquatica*	NEB 195108
	Catalpa speciosa	*Catalpa speciosa*	In the *Report* but not the *List*
38	*Celastrus scandens*	*Celastrus scandens*	NEB 179507
364	*Celtis occidentalis*	*Celtis occidentalis*	No specimen at NEB
411	*Cenchrus tribuloides*	*Cenchrus longispinus*	No specimen at NEB
92	*Cercocarpus parvifolius*	*Cercocarpus montanus*	NEB 175025; RM 224075
199	*Chaenactis douglasii* var. *alpina*	*Chaenactis douglasii* var. *alpina*	NEB 188970, 191488
s.n.	*Chara* sp.	*Chara foetida*	NEB 227432
479	*Cheilanthes lanuginosa*	*Cheilanthes feei*	NEB 167678, 167679, 167680, 167681
322	*Chenopodium album*	*Chenopodium berlandieri*	NEB 172173, 172190, 172198
317	*Chenopodium fremontii* var. *incanum*	*Chenopodium incanum*	NEB 172312
318	*Chenopodium leptophyllum*	*Chenopodium pratericola*	NEB 172345, 172381
319	*Chenopodium leptophyllum* var. *oblongifolium*	*Chenopodium desiccatum*	NEB 172347
320	*Chenopodium leptophyllum* var. *subglabrum*	*Chenopodium pratericola*	NEB 172362
321	*Chenopodium* sp.	*Chenopodium fremontii*	NEB 172248, 172258, 172259
146	*Chrysopis* sic *villosa*	*Heterotheca villosa*	NEB 188875, 207821
147	*Chrysopis* sic *villosa* var. *hispida*	*Heterotheca villosa*	NEB 131631, 188804
417	*Chrysopogon nutans*	*Sorghastrum nutans*	NEB 197760
149 *sic*	*Chrysopsis villosa* var. *canescens*	*Heterotheca villosa*	NEB 188874
149	*Chrysopsis villosa* var. *sessiliflora*	*Heterotheca villosa*	NEB 188813
131	*Cicuta maculata*	*Cicuta maculata*	NEB 181567
1	*Clematis ligusticifolia*	*Clematis ligusticifolia*	NEB 173367, 173391
27	*Cleome angustifolia*	*Cleomella angustifolia*	NEB 174695, 174697, 174705
26	*Cleome integrifolia*	*Cleome serrulata*	NEB 174656
216	*Cnicus altissimus* var. *filipendulum*	*Cirsium flodmanii*	NEB 198233; NY 2051752, 2051799
214	*Cnicus ochrocentrus*	*Cirsium ochrocentrum*	NEB 1889144, 188945; NY 2052626
213	*Cnicus pitcheri*	*Cirsium undulatum*	NEB 188918
215	*Cnicus undulatus*	*Cirsium undulatum*	NEB 188923; NY 2053552
262	*Convolvulus sepium*	*Calystegia sepium*	No specimen at NEB
328	*Copiospermum (sic) hyssopifolium*	*Corispermum americanum*	NEB 172415
189	*Coreopsis tinctoria*	*Coreopsis tinctoria*	NEB 189121
132	*Cornus stolonifera*	*Cornus sericea*	No specimen at NEB
10	*Corydalis curvisiliqua*	*Corydalis curvisiliqua* ssp. *occidentalis*	NEB 173939
217	*Crepis runcinata*	*Crepis runcinata*	NEB 187251, 187257, 207819
363	*Croton texensis*	*Croton texensis*	NEB 178528; US 102258
219 *sic*	*Cucurbita foetidissima*	*Cucurbita foetidissima*	NEB 186919
264	*Cuscuta* sp.	*Cuscuta* sp.	No specimen at NEB
316	*Cycloloma platyphyllum*	*Cycloloma atriplicifolium*	NEB 172429
130	*Cymopteris montanus*	*Cymopteris montanus*	No specimen at NEB
391	*Cyperus aristatus*	*Cyperus squarrosus*	NEB 192906, 193053
392	*Cyperus schweinitzii*	*Cyperus schweinitzii*	NEB 193098, 193103
54	*Dalea alopecuroides*	*Dalea leporina*	NEB 178208; NY 1270107
55	*Dalea laxiflora*	*Dalea enneandra*	NY 1259327
8	*Delphinium azureum*	*Delphinium virescens*	NEB 173451
88	*Desmanthus brachylobus*	*Desmanthus illinoensis*	NEB 176033, 176034

Rydberg's number	Name used by Rydberg	Modern or corrected name	Vouchers
84	*Desmodium canadense*	*Desmodium canadense*	NEB 178224, 178237
455	*Distichlis spicata* var. *stricta*	*Distichlis spicata*	NEB 195261
224	*Dodecantheon (sic) meadia*	*Dodecatheon pulchellum*	No specimen at NEB
11	*Draba carolin(ian)a* var. *micrantha*	*Draba reptans*	NEB 174390
200	*Dys(s)sodia chrysanthemoides*	*Dyssodia papposa*	NEB 189182
452	*Eatonia obtusata*	*Sphenopholis obtusata*	No specimen at NEB
453	*Eatonia* sp.	*Sphenopholis obtusata*	NEB 195334
249	*Echinospermum floribundum*	*Hackelia floribunda*	NEB 183679
251	*Echinospermum redowskii* var. *cupulatum*	*Lappula redowskii* var. *cupulata*	NEB 183722, 207847
250	*Echinospermum redowskii* var. *occidentale*	*Lappula redowskii* var. *redowskii*	NEB 183700
398	*Eliocharis (sic)* sp.	*Eleocharis compressa*	NEB 193258, 193269
397	*Eliocharis (sic) acicularis*	*Eleocharis acicularis*	NEB 193231, 193258
473	*Elymus canadensis*	*Elymus canadensis*	NY 1654755, 1654759; RM 2378
472	*Elymus sitanion*	*Elymus elymoides* subsp. *brevifolius*	NEB 197734
107	*Epilopbium adenocaulon*	*Epilobium ciliatum*	NEB 180916, 180920
476	*Equisetum laevigatum*	*Equisetum laevigatum*	NEB 167973, 167977
475	*Equisetum limosum*	*Equisetum laevigatum*	NEB 167934, 167998
477	*Equisetum variegatum*	*Equisetum laevigatum*	NEB 168028
172	*Erigeron canadensis*	*Conyza canadensis*	No specimen at NEB
169	*Erigeron canus*	*Erigeron canus*	NEB 189356
170	*Erigeron divergens*	*Erigeron bellidiastrum*	NEB 189427, 207809
168	*Erigeron pumilus*	*Erigeron pumilus*	NEB 189412, 207833
171	*Erigeron strigosus*	*Erigeron strigosus*	NEB 189448, 189464, 189466
330	*Eriogonum alatum*	*Eriogonum alatum* var. *alatum*	NEB 171131, RM 2557
332	*Eriogonum annum (sic)*	*Eriogonum annuum*	NEB 171088
338	*Eriogonum brevicaule*	*Eriogonum brevicaule*	NEB 171137, 207860
333	*Eriogonum cernuum*	*Eriogonum cernuum*	NEB 171142
337	*Eriogonum corymbosum*	*Eriogonum nebraskense*	NEB 171172; NY 323405, 323406; US 63360
331	*Eriogonum flavum*	*Eriogonum flavum* var. *flavum*	NEB 171154, 171164
335	*Eriogonum microthecum*	*Eriogonum effusum* var. *effusum*	NEB 171177
336	*Eriogonum microthecum* v. *effusum*	*Eriogonum effusum* var. *effusum*	NEB 171171, 171172; RM 225453
334	*Eriogonum multiceps*	*Eriogonum pauciflorum* var. *gnaphalodes*	NEB 171192
16	*Erysimum asperum*	*Erysimum asperum*	NEB 174163
15	*Erysimum cheiranthoides*	*Erysimum cheiranthoides*	No specimen at NEB
233	*Erythraea douglasii*	*Centaurium exaltatum*	NEB 182787
361	*Euphorbia cuphusperma (sic)*	*Euphorbia davidii*	US 91611
358	*Euphorbia fendleri*	*Euphorbia fendleri*	NEB 178670
355	*Euphorbia glyptosperma*	*Euphorbia glyptosperma*	NEB 178700
359	*Euphorbia hexagona*	*Euphorbia hexagona*	NEB 178763
357	*Euphorbia maculata*	*Euphorbia maculata*	No specimen at NEB
360	*Euphorbia marginata*	*Euphorbia marginata*	No specimen at NEB
362	*Euphorbia montana*	*Euphorbia robusta*	NEB 178840
353	*Euphorbia petaloidea*	*Euphorbia missurica* var. *petaloidea*	NEB 207859
354	*Euphorbia serpyllifolia*	*Euphorbia serpyllifolia*	No specimen at NEB
356	*Euphorbia* sp.	*Euphorbia serpyllifolia*	NY 263066; US 65781
327	*Eurotia lanata*	*Krascheninnikovia lanata*	NEB 172488, 172491, 172494

Rydberg's number	Name used by Rydberg	Modern or corrected name	Vouchers
234	*Eustoma russellianum*	*Eustoma grandiflorum*	NEB 182424
263	*Evolvulus argenteus*	*Evolvulus nuttallianus*	NEB 183041
462	*Festuca tenella*	*Vulpia octoflora*	NEB 196068
399	*Fimbristylis spadicea* var. *castanea*	*Fimbristylis puberula*	NEB 193473, 193475
180	*Franseria discolor*	*Ambrosia tomentosa*	NEB 189643, 189645, 207898; RM 2647
179	*Franseria hookeriana*	*Ambrosia acanthicarpa*	NEB 189654, 189657, 189659, 189660; NY 187754, 187766; RM 2647
181	*Franseria tomentosa*	*Ambrosia grayi*	NEB 133898, 189664
225	*Fraxinus pubescens*	*Fraxinus pennsylvanica*	NEB 182320, 207841
257	*Fraxinus pubescens*	probably *F. pennsylvanica*	No specimen at NEB
376	*Fritellaria atro-purpurea (sic)*	*Fritillaria atropurpurea*	NEB 169538
314	*Froelichia floridana*	*Froelichia floridana*	NEB 1729049
315	*Froelichia gracilis*	*Froelichia gracilis*	NEB 172077, 172079
134	*Galium boreale*	*Galium boreale*	NEB 181531, 186541, 186543, 207811, 207812; NY 131422
118	*Gaura coccinea*	*Gaura coccinea*	NEB 181010, 181018, 181039
119	*Gaura coccinea* var. ___	*Gaura coccinea*	NEB 181031
120	*Gaura parviflora*	*Gaura parviflora*	No specimen at NEB
284	*Gerardia purpurea*	*Agalinis purpurea*	No specimen at NEB
285	*Gerardia tenuifolia*	*Agalinis tenuifolia* var. *parviflora*	No specimen at NEB
93	*Geum album*	*Geum canadense*	No specimen at NEB
94	*Geum strictum*	*Geum aleppicum*	NEB 175202, 175205
241	*Gilia gracilis*	*Microsteris gracilis* var. *humilior*	NEB 183305, 207837
245	*Gilia iberidifolia*	*Ipomopsis congesta*	NEB 183245, 183284, 183293
240	*Gilia linearis*	*Collomia linearis*	NEB 183305, 207837; NY 2361314
242	*Gilia longiflora*	*Ipomopsis longiflora*	NEB 183335, 183345
243	*Gilia pungens* var. *caespitosa*	*Leptodactylon caespitosum*	NEB 183285, 183286
246	*Gilia* sp.	*Gilia pinnatifida*	NEB 183351
244	*Gilia spicata*	*Ipomopsis spicata*	NEB 183354, 207835
459	*Glyceria distans*	*Puccinellia nuttalliana*	NEB 197519
460	*Glyceria nervata*	*Glyceria striata*	No specimen at NEB
461	*Glyceria sp.*	*Poa glaucifolia*	NY 431306
s.n.	*Glycyrrhiza lepidota*	*Glycyrrhiza lepidota*	In the *Report* and *List;* no specimen at NEB
141	*Grindelia squarrosa*	*Grindelia squarrosa*	NEB 189688
140	*Gutierrezia euthamiae*	*Gutierrezia sarothrae*	NEB 189729, 189742, 207814
297	*Hedeoma drummondii*	*Hedeoma drummondii*	NEB 184350, 207857
296	*Hedeoma hispida*	*Hedeoma hispida*	No specimen at NEB
197	*Helenium autumnale*	*Helenium autumnale*	NEB 189751, 189765, 190681, 207823
184	*Helianthus annus (sic)*	*Helianthus annuus*	No specimen at NEB
187	*Helianthus maximiliani*	*Helianthus maximiliani*	NEB 199111
188	*Helianthus maximiliani* var. ___	*Helianthus maximiliani*	No specimen at NEB
185	*Helianthus petiolaris*	*Helianthus petiolaris*	NEB 189927
186	*Helianthus rigidus*	*Helianthus pauciflorus* subsp. *subrhomboideus*	NEB 189960
248	*Heliotropium currasivicum (sic)*	*Heliotropium curassavicum* var. *obovatum*	NEB 183651
47	*Hosackia purshiana*	*Lotus purshianus, L. unifoliatus, Acmispon americanus*	NEB 177020, 177025

Rydberg's number	Name used by Rydberg	Modern or corrected name	Vouchers
195	*Hymenopappus* sp.	*Hymenopappus filifolius* var. *polycephalus*	NEB 189819, 189854
194	*Hymenopappus tenuifolius* var.	*Hymenopappus tenuifolius*	NEB 189832
261	*Ipomaea (sic) leptophylla*	*Ipomoea leptophylla*	NEB 183087, 207850
175	*Iva axillaris*	*Iva axillaris*	NEB 190499, 190500
174	*Iva ciliata*	*Iva annua*	NEB 190516, 207807
176	*Iva xanthifolia*	*Cyclachaena xanthiifolia*	NEB 190533
	Juglans nigra	*Juglans nigra*	In the *Report* but not the *List*
380	*Juncus bufonius*	*Juncus bufonius*	NEB 155528, 168982, 168992, 168994
378	*Juncus filiformis*	Probably *Juncus arcticus* var. *balticus*	No specimen at NEB
382	*Juncus nodosus* var. ___	*Juncus nodosus*	No specimen at NEB
381	*Juncus nodosus* var. *megacephalus*	*Juncus nodosus*	No specimen at NEB
379	*Juncus tenuis*	*Juncus tenuis*	No specimen at NEB
474	*Juniperus virginiana*	*Juniperus scopulorum*	NY 20817
	Juniperus virginiana	*Juniperus virginiana*	In the *Report* but not the *List*
450	*Koeleria cristata*	*Koeleria macrantha*	NEB 196248
451	*Koeleria cristata* var. *gracilis*	*Koeleria macrantha?*	No specimen at NEB
252	*Krynitzkia crassisepala*	*Cryptantha minima*	NEB 183616, 183621
253	*Krynitzkia fendleri*	*Cryptantha fendleri*	NEB 183598, 183612
255	*Krynitzkia glomerata*	*Cryptantha thyrsiflora*	NEB 114094, 184004, 207848, 207849; RM 82748s
254	*Krynitzkia jamesii*	*Cryptantha cinerea* var. *jamesii*	NEB 183971, 183977, 183988, 183989, 207847, 207923
256	*Krynitzkia sericea*	*Cryptantha cana*	NEB 183944
136	*Kuhnia eupatorioides* var. *corymbulosa*	*Brickellia eupatorioides* var. *corymbulosa*	NEB 190011
218	*Lactuca ludoviciana*	*Lactuca ludoviciana*	NEB 187321
86	*Lathyrus ornatus*	*Lathyrus decaphyllus*	NEB 177256
413	*Leersia oryzoides*	*Leersia oryzoides*	NEB 196147
385	*Lemna minor*	*Lemna* ref. *minor*	NEB 248587
386	*Lemna perpusilla*	*Lemna perpusilla*	NEB 248588
182	*Lepachys columnaris*	*Ratibida columnifera*	NEB 190437
183	*Lepachys columnaris* var. *pulcherrima*	*Ratibida columnifera* f. *pulcherrima*	NEB 190402, 190404
	Lespedeza capitata	*Lespedeza capitata*	In the *Report* but not the *List*
22	*Lesquerella engelmannii*	*L. ovalifolia, Physaria ovalifolia* var. *ovalifolia*	NEB 77288; NY 185553, 185554, US 4849
23	*Lesquerella ludoviciana*	*Physaria ludoviciana*	NEB 77271,77272
138	*Liatris punctata*	*Liatris punctata*	NEB 190248
139	*Liatris spicata*	*Liatris lancifolia*	NEB 190341
281	*Limosella aquatica*	*Limosella aquatica*	NEB 185655
37	*Linum rigidum*	*Linum compactum*	NEB 179179
306	*Lippia cuneifolia*	*Lippia cuneifolia*	NEB 184027, 184048
259	*Lithospermum angustifolium*	*Lithospermum incisum*	NEB 183873
258	*Lithospermum hirtum*	*Lithospermum carolinense* (?)	No specimen at NEB
222	*Lobelia spicata*	*Lobelia spicata*	NEB 187121, 207443, 207844; US 57811
45	*Lupinus argenteus* var. *decumbens*	*Lupinus argenteus* vars. *argenteus* and *argophyllus*	NEB 177381, 177388, 177379. 177380; NY 2587650
46	*Lupinus pusillus*	*Lupinus pusillus*	NEB 177487, 311890
294	*Lycopus lucidus* var. *americanus*	*Lycopus americanus*	No specimen at NEB

Rydberg's number	Name used by Rydberg	Modern or corrected name	Vouchers
295	*Lycopus sinuatus*	?	No specimen at NEB
220	*Lygodesmia juncea*	*Lygodesmia juncea*	NEB 187446
36	*Malvastrum coccineum*	*Sphaeralcea coccinea*	NEB 180102
s.n.	*Mammil(l)ari(a) vivipara*	*Coryphantha vivipara*	NEB 180531
	Medicago sativa	*Medicago sativa*	In the *Report* but not the *List*
293	*Mentha canadensis*	*Mentha arvensis*	NEB 184607
123	*Mentzelia nuda*	*Mentzelia nuda*	NEB 180407, 180471
122	*Mentzelia ornata*	*Mentzelia decapetala*	NEB 180443
121	*Mentzelia sp.*	*Mentzelia albicaulis*	NEB 180076, 180437, 180438, 180439
65	*Millelotus (sic) alba*	*Melilotus albus*	NEB 178292, 178293
280	*Mimulus jamesii*	*Mimulus glabratus* var. *jamesii*	NEB 185697
311	*Mollugo verticillata*	*Mollugo verticillata*	No specimen at NEB
301	*Monarda citriodora*	*Monarda pectinata*	NEB 184653, 184712, 207858
299	*Monarda fistulosa*	*Monarda fistulosa* var. *menthifolia*	NEB 184656
300	*Monarda fistulosa* var. *mollis*	*Monarda fistulosa* var. *menthifolia*	NEB 184637, 184646, 184858
323	*Monolepis chenopodioides*	*Monolepis nuttalliana*	NEB 172559, NY 539850
	Morus alba	*Morus alba*	In the *Report* but not the *List*
429	*Muhlenbergia pungens*	*Muhlenbergia pungens*	NY 1741003
428	*Muhlenberiga ramosa*	*Muhlenbergia racemosa*	NEB 196540
449	*Munroa squarrosa*	*Munroa squarrosa*	No specimen at NEB
127	*Musineum (sic) tenuifolium*	*Musineon tenuifolium*	NEB 181628, 181734
20	*Nasturtium curvisiliqua*	*Rorippa sinuata*	NEB 177487
18	*Nasturtium obtusum*	*Rorippa tenerrima*	NEB 174451
19	*Nasturtium palustre*	*Rorippa palustris*	NEB 174470
21	*Nasturtium sinuatum*	*Rorippa sinuata*	NEB 174417, 174428
40	*Negundo aceroides*	*Acer negundo*	No specimen at NEB
111	*Oenothera albicaulis*	*Oenothera nuttallii, O. albicaulis* var. *nuttallii*	NEB 180820, 180829; US 44483
112	*Oenothera albicaulis* var. ___	*Oenothera pallida* subsp. *latifolia*	US 44482
108	*Oenothera biennis* var. ___	*Oenothera biennis*	NEB 181216
114	*Oenothera caespitosa* var. ___	*Oenothera caespitosa*	NEB 181385, US 55901
115	*Oenothera canescens*	*Oenothera canescens*	NEB 181117, US 56677
113	*Oenothera coronopifolia*	*Oenothera coronopifolia*	NEB 180770, US 44484
117	*Oenothera hartwegii* var. *lavandulaefolia*	*Oenothera lavandulifolia, Calylophus lavandulifolius*	NEB 180982, US 44060
110	*Oenothera pinnatifida*	*Oenothera albicaulis* var. *nuttallii*	NEB 180851, 180852
116	*Oenothera serrulata*	*Oenothera serrulata*	NEB 181331
109	*Oenothera sinuata*	*Oenothera laciniata*	NEB 181270, US 56675
260	*Onosmodium carolinianum*	*Onosmodium molle* var. *occidentale*	NY 335611
126	*Opuntia fragilis*	*Opuntia fragilis*	NEB 180497; US 3046541, 3046821
124	*Opuntia missouriensis*	*Opuntia polyacantha* var. *polyacantha*	NEB 180524; US 3046542
125	*Opuntia rutila*	*Opuntia polyacantha* var. *polyacantha*	NEB 180494, 180521, 180526, 180527; US 3046540
287	*Orthocarpus luteus*	*Orthocarpus luteus*	NEB 120251, 185809, 185811
427	*Oryzopsis cuspidata*	*Stipa hymenoides*	RM 224521
426	*Oryzopsis micrantha*	*Oryzopsis micrantha*	RM 223522

Rydberg's number	Name used by Rydberg	Modern or corrected name	Vouchers
308	*Oxybaphus angustifolius*	*Mirabilis linearis*	NEB 172678
307	*Oxybaphus nyctaginea* var. *oblongifolius*	*Mirabilis hirsuta*	NEB 172722
82	*Oxytropis*	*Oxytropis multiceps*	NEB 177742
82b	*Oxytropis lambertii*	*Oxytropis lambertii*	NEB 176174, 176176; NY 5224
83	*Oxytropis lambertii* var. *sericea*	*Oxytropis sericea*	NEB 177734
405	*Panicum capillare*	*Panicum capillare*	No specimen at NEB
410	*Panicum crus-galli*	Probably *Echinochloa muricata* var. *microstachya*	No specimen at NEB
409	*Panicum dichotomum* var. *barbulatum*	Perhaps *Panicum acuminatum*	No specimen at NEB
408	*Panicum scoparium*	Perhaps *Panicum leibergii*	No specimen at NEB
406	*Panicum virgatum*	*Panicum virgatum*	No specimen at NEB
407	*Panicum virgatum* var. *confertum*	*Panicum virgatum*	No specimen at NEB
366	*Parietaria pensylvanica*	*Parietaria pensylvanica*	RM 248313
310	*Paronychia jamesii*	*Paronychia jamesii*	NEB 172942
201	*Pectis angustifolia*	*Pectis angustifolia*	NEB 190660, 190661, 190662, 207825
275	*Penstemon frimontii (sic)*	*Penstemon albidus*	NY 541697
276	*Penstemon glaber*	*Penstemon glaber* var. *glaber*	NEB 185933, 185934; NY 595804
279	*Penstemon gracilis*	*Penstemon glaber* var. *gracilis*	NEB 186060, 207855; NY 600776, 600850
277	*Penstemon grandiflorus*	*Penstemon grandiflorus*	NEB NY 601122, RM 2515
278	*Penstemon humilis*	*Penstemon albidus*	NEB 185847; NY 541687, 541698
58	*Petalostemon*	*Dalea candida* var. *oligophylla*	NEB 177244, 177248; NY 1241329, 1241339; RM 250565
57	*Petalostemon candidus*	*Dalea candida* var. ___	No specimen at NEB
	Petalostemon gracile	*Dalea candida* var. oligophylla	In the *Report* but not the *List*
61	*Petalostemon macrostachyus*	*Dalea cylindriceps*	NEB 177086, 177088; NY 1259216, 1259217
59	*Petalostemon tenuifolius*	*Dalea purpurea* var. *purpurea and* var. *arenicola*	NY 1277408, 1277410, 1277675.
62	*Petalostemon villosus*	*Dalea villosa*	NEB 177210, NY 1277877
60	*Petalostemon violaceum*	*Dalea purpurea* var. *purpurea*	No specimen at NEB
247	*Phacelia* sp.	*Phacelia hastata*	NEB 183511, 183555, 207886; RM 238975
419	*Phalaris arundinacea*	*Phalaris arundinacea*	NEB 197171
418	*Phalaris canariensis*	*Phalaris canariensis*	NEB 197191, 197192
237	*Phlox caespitosa* var. *rigida*	*Phlox hoodii*	No specimen at NEB
238	*Phlox douglasii* var.	*Phlox andicola*	NEB 183387, 207839, 207840,
239	*Phlox douglasii* var. *longifolia*	*Phlox andicola*	No specimen at NEB
236	*Phlox hoodii*	*Phlox hoodii*	NEB 183436
235	*Phlox muscoides*	*Phlox bryoides*	NEB 183397, NY 336979
448	*Phragmites communis*	*Phragmites australis*	No specimen at NEB
274	*Physalis lanceolata*	*Physalis virginiana*	NEB 185326
272	*Physalis longifolia*	*Physalis longifolia*	NEB 185254, 185269, 207854
269	*Physalis mollis* var. *cinerascens*	*Physalis hederifolia* var. *comata*, *Physalis comata*	NEB 185153; NY 138822, 138823; US 126246
270	*Physalis* sp.	*Physalis hederifolia* var. *comata*	NEB 185282
271	*Physalis* sp.	*Physalis longifolia*	NEB 185231, 185267
273	*Physalis* sp.	*Physalis hispida*	NEB 185234, 185243; RM 2503
268	*Physalis virginana*	*Physalis heterophylla*	NEB 181565. 185177

Rydberg's number	Name used by Rydberg	Modern or corrected name	Vouchers
24	*Physaria didymocarpa*	*Physaria brassicoides*	NEB 77497, 174378; NY 185608; RM 2615; US 4739
303	*Physostegia parviflora*	*Physostegia parviflora*	No specimen at NEB
	Pinus ponderosa var. *scopulorum*	*Pinus ponderosa* var. *scopulorum*	In the *Report* but not the *List*
	Plantago lanceolata	*Plantago lanceolata*	In the *Report* but not the *List*
305	*Plantago patagonica* var. *gnaphalioides (sic)*	*Plantago patagonica*	NEB 186427, 186435
304	*Plantago sp.*	*Plantago eriopoda*	NEB 136370, 186369
367	*Platanus occidentalis*	*Platanus occidentalis*	NEB 174650
456	*Poa alpina*	*Poa pratensis*	NEB 197270
458	*Poa nevadensis*	*Poa secunda*	NEB 197500
461	*Poa plattensis*	*Poa glaucifolia*	NY 431306
457	*Poa tenuifolia*	*Poa arida*	NEB 197499; NY 1784748, 1784752, 1784753
25	*Polanisia trachysperma*	*Polanisia dodecandra* subsp. *trachysperma*	NEB 174733
29	*Polygala alba*	*Polygala alba*	No specimen at NEB
350	*Polygonum convolvulus*	*Polygonum convolvulus*	NEB 171276
351	*Polygonum dumetorum* v. *scandens*	*Polygonum scandens*	NEB 171628, 171633
343	*Polygonum erectum*	*Polygonum buxiforme*	NEB 171429
347	*Polygonum lapatifolium (sic)*	*Polygonum lapathifolium*	NEB 171403
348	*Polygonum lapatifolium (sic)* var. *incanum*	*Polygonum lapathifolium*	No specimen at NEB
349	*Polygonum muhlenbergii*	*Polygonum coccineum, Persicaria amphibia (sens. lat.)*	NEB 171288
345	*Polygonum ramosissimum*	*Polygonum ramosissimum*	NEB 171262, 171516
344	*Polygonum sp.*	*Polygonum aviculare*	NEB 171240, 171429
346	*Polygonum tenue*	Probably *Polygonum sawatchense*	No specimen at NEB
	Populus deltoides	*Populus deltoides* subsp. *monilifera*	In the *Report* but not the *List*
372	*Populus* sp.	*Populus* x *acuminata*	GH 30820; NY 283994; US 15164
	Portulacca (sic) oleracea	*Portulaca oleracea*	In the *Report* but not the *List*
97	*Potentilla anserina*	*Potentilla anserina*	NEB 175019
96	*Potentilla pensylvanica*	*Potentilla pensylvanica*	No specimen at NEB
95	*Potentilla supina*	*Potentilla paradoxa*	NEB 175330, 175202
	Prunus americana	*Prunus americana*	In the *Report* but not the *List*
90	*Prunus demissa*	*Prunus virginiana*	No specimen at NEB
91	*Prunus pumila*	*Prunus pumila* var. *besseyi*	NEB 175883
129	*Pseudocymopterus anisatus*	*Lomatium nuttallii*	NEB 181619; NY 1141437
49	*Psoralea argophylla*	*Pediomelum argophyllum*	No specimen at NEB
50	*Psoralea argophylla* var.___	*Pediomelum argophyllum*	NEB 177854; NY 26763; US 24014
51	*Psoralea campestris*	*Pediomelum digitatum*	NEB 178047, 178055, 177989, 178037, 178052
	Psoralea esculenta	*Pediomelum esculentum*	In the *Report* but not the *List*
52	*Psoralea hypogaea*	*Pediomelum hypogaeum*	NEB 177988
53	*Psoralea lanceolata*	*Psoralidium lanceolatum*	No specimen at NEB
48	*Psoralea tenuiflora*	*Psoralidium tenuiflorum*	NEB 178104
4	*Ranunculus circinatus*	*Ranunculus longirostris*	NEB 173315
5	*Ranunculus cymbalaria*	*Ranunculus cymbalaria*	NEB 173648
7	*Ranunculus hispidus*	*Ranunculus macounii*	NEB 173710
6	*Ranunculus sceleratus*	*Ranunculus sceleratus*	NEB 173724
42	*Rhus canadensis* var. *trilobata*	*Rhus aromatica*	NEB 179148

Rydberg's number	Name used by Rydberg	Modern or corrected name	Vouchers
41	*Rhus toxicodendron*	*Toxicodendron radicans* var. *rydbergii*	NEB 179110
106	*Ribes aureum*	*Ribes odoratum*	NEB 79514; US 590343, 691782
104	*Ribes cereum*	*Ribes cereum* var. *inebrians*	No specimen at NEB
105	*Ribes floridum*	*Ribes americanum*	No specimen at NEB
	Ribes gracile	*Ribes missouriense*	In the *Report* but not the *List*
99	*Rosa arkansana*	*Rosa arkansana*	NEB 175520
98	*Rosa blanda*	*Rosa woodsii*	NEB 175426
101	*Rosa fendleri*	*Rosa woodsii*	NEB 175461
100	*Rosa nutkana*	*Rosa woodsii*	NEB 175453
102	*Rosa woodsii*	*Rosa woodsii*	NEB 175476; RM 224161A, 224161B
342	*Rumex maritimus*	*Rumex fueginus*	NEB 171781; NY 2405694, 2408702
340	*Rumex salicifolius*	*Rumex triangulivalvis*	NEB 171834; NY 2409549
341	*Rumex* sp.	*Rumex crispus*	NY 2407781
339	*Rumex venosus*	*Rumex venosus*	NEB 171852; NY 2409761, 2409780
388	*Sagittaria variabilis*	*Sagittaria cuneata*	NEB 168775
	Salix amygdaloides	*Salix amygdaloides*	In the *Report* but not the *List*
371	*Salix cordata* var. *vestita*	*Salix amygdaloides*	NY 2404372, 2404375
368	*Salix longifolia*	*Salix interior, S. exigua* subsp. *interior*	NEB 170257; NY 73754; RM 2580
369	*Salix longifolia* var. ___	*Salix interior, S. exigua* subsp. interior	NY 2415092
370	*Salix longifolia* var. ___	*Salix interior, S. exigua* subsp. ___	No specimen at NEB
298	*Salvia lanceolata*	*Salvia reflexa*	NEB 184821
443	*Schedon(n)ardus texanus*	*Schedonnardus paniculatus*	NEB 197581
89	*Schrankia uncinata*	*Mimosa quadrivalvis* var. *nuttallii*	NEB 176068, 176078
396	*Scirpus atrovirens* var. *pallidus*	*Scirpus pallidus*	NEB 193660
394	*Scirpus lacustris*	*Schoenoplectus tabernaemontani*	NEB 193720, 193725
395	*Scirpus maritimus*	*Bolboschoenus fluviatilis*	NEB 193607
393	*Scirpus pungens*	*Schoenoplectus pungens*	NEB 19559, 193560
302	*Scutellaria galericulata*	*Scutellaria galericulata*	NEB 184902
211	*Senecio aureus* var. ___	*Packera plattensis*	NEB 191501, 207818
210	*Senecio canus*	*Packera cana*	NEB 191415
212	*Senecio douglasii*	*Senecio riddellii*	NEB 191589
352	*Shepherdia argentea*	*Shepherdia argentea*	NEB 180563
14	*Sinapis arvensis*	*Sinapis arvensis*	No specimen at NEB
17	*Sisymbrium canescens*	*Descurainia pinnata*	NEB 174569
373	*Sisyrinchium anceps*	*Sisyrinchium montanum*	NEB 169137
	Smilacina stellata	*Maianthemum stellatum*	In the *Report* but not the *List*
266	*Solanum nigrum*	*Solanum interius*	No specimen at NEB
267	*Solanum rostratum*	*Solanum rostratum*	NEB 185452, 207853; NY 820854, 820855
265	*Solanum triflorum*	*Solanum triflorum*	NEB 18546…; NY 820258, 820284
155	*Solidago canadensis*	*Solidago canadensis*	NEB 207830
159	*Solidago lanceolata*	*Euthamia gymnospermoides*	NEB 189560
152	*Solidago missouriensis*	*Solidago missouriensis*	NEB 191006, 190966, 190975, 191004, 191035; NY 2354482
153	*Solidago missouriensis*	*Solidago missouriensis*	NEB 190969, 207831

Rydberg's number	Name used by Rydberg	Modern or corrected name	Vouchers
154	*Solidago missouriensis* var. ___	*Solidago missouriensis*	NEB 207829
158	*Solidago nemoralis* var. ___	*Solidago mollis*	NEB 191069, 191073
157	*Solidago nemoralis* var. *incana*	*Solidago mollis*	NEB 191068
156	*Solidago serotina*	*Solidago gigantea*	NEB 191211
44	*Sophora sericea*	*Sophora nuttalliana*	NEB 177659, 177662, 177668
384	*Sparganium eurycarpum*	*Sparganium eurycarpum*	NEB 168306, 168307
412	*Spartina cynocurodes (sic)*	*Spartina pectinatus*	No specimen at NEB
434	*Sporobolus airoides*	*Sporobolus airoides*	NY 1819445; RM 2602, 22172
435	*Sporobolus asperifolius*	*Muhlenbergia asperifolia*	NY 1740431, 1740433
432	*Sporobolus cryptandrus*	*Sporobolus cryptandrus*	No specimen at NEB
433	*Sporobolus cryptandrus* var. ___	*Sporobolus cryptandrus*	NEB 197951; RM 222191
430	*Sporobolus cuspidatus*	*Muhlenbergia cuspidata*	NY 1755928, 1755974
431	*Sporobolus cuspidatus* var. ___	*Muhlenbergia cuspidata*	NEB 196396
436	*Sporobouls serotinus* var. ___	*Muhlenbergia minutissima*	NEB 312300
219	*Stephanomeria minor*	*Stephanomeria runcinata*	NEB 187576, 187577, 187666, 207822; NY 114892, 114893; RM 97231s; US 75555
424	*Stipa comata*	*Stipa comata*	NEB 198113
425	*Stipa viridula*	*Nasella viridula*	NEB 198254
87	*Strophostyles pauciflorus*	*Strophostyles leiosperma*	NEB 177693
329	*Suaeda diffusa*	*Suaeda calceoliformis*	NEB 172454
133	*Symphoricarpos occidentalis*	*Symphoricarpos occidentalis*	No specimen at NEB
	Symphoricarpos orbiculatus	*Symphoricarpos orbiculatus*	In the *Report* but not the *List*
33	*Talinum teretifolium*	*Phemeranthus parviflorus*	RM 250536
292	*Teucrium occidentale*	*Teucrium canadense* var. *occidentale*	No specimen at NEB
3	*Thalictrum purpurascens*	*Thalictrum dasycarpum*	NEB 173788, 173807
192	*Thelesperma filifolium*	*Thelesperma filifolium* var. *intermedium*	NEB 139976, 139980, 207824; US 6114, 76488
193	*Thelesperma gracilis*	*Thelesperma megapotamicum*	NEB 140013
43	*Thermopsis rhombifolia*	*Thermopsis rhombifolia*	NEB 91413
160	*Townsendia grandiflorus*	*Townsendia grandiflorus*	NEB 207832
390	*Triglochin maritima*	*Triglochin maritima*	NEB 168627, 168641
383	*Typha latifolia*	*Typha latifolia*	No specimen at NEB
	Ulmus americana	*Ulmus americana*	In the *Report* but not the *List*
	Ulmus rubra	*Ulmus rubra*	In the *Report* but not the *List*
365	*Urtica gracilis*	*Urtica dioica* subsp. *gracilis*	No specimen at NEB
135	*Vernonia fasciculata*	*Vernonia fasciculata*	NEB 191663
282	*Veronica americana*	*Veronica americana*	NEB 186140, 186146; RM 2517
283	*Veronica peregrina* var.	*Veronica peregrina* var. *xalapensis*	No specimen at NEB
85	*Vic(i)a americana* var. *linearis*	*Vicia americana* var. *minor*	NEB 207881
28	*Viola nuttallii*	*Viola nuttallii*	NEB 180291
s.n.	*Vitis riparia*	*Vitis riparia*	In the *Report* and *List,* and perhaps not collected. No specimen at NEB.
375	*Yucca angustifolia*	*Yucca glauca*	NEB 169520
389	*Zannichellia palustris*	*Zannichellia palustris*	NEB 168618, 168619
377	*Zygadenus* sp.	*Zigadenus venenosus, Toxicoscordion gramineum*	NEB 169538

Literature Cited

Barnhart, J. H. 1931. Per Axel Rydberg. *Journal of the New York Botanical Garden* 32: 229–233.

Bessey, Charles E. 1892. *Native trees and shrubs of Nebraska.* Annual Report of the Nebraska State Horticultural Society, 1892.

Coulter, John M. 1885. *Coulter's Manual of Rocky Mountain Botany.* New York & Chicago: Ivison, Blakeman, Taylor and Company. 480 pp.

Flora of North America Editorial Committee. 1993–. *Flora of North America, North of Mexico,* multiple volumes. Oxford University Press.

Fuenning, Samuel L. 1960. Walter Bigler Kiener (1894-1959). *The Bryologist* 63: 64–66.

Gray, Asa, et al. 1878-. *Synoptical Flora of North America.* Vols. 1 & 2 + supplements in 1886, 1888. New York: Ivison, Blakeman, Taylor and Company; American Book Co.; Smithsonian Miscellaneous Collections 491, Smithsonian Institution, Washington DC.

Harris, Earl R. 1962. Courthouse and Jail Rocks: Landmarks on the Oregon Trail. *Nebraska History* 43: 29–51 + 5 plates.

Kaul, Robert B., David M. Sutherland, Steven B. Rolfsmeier. 2011. *The Flora of Nebraska,* second edition. Lincoln: University of Nebraska-Lincoln, School of Natural Resources. 967 pp.

Kiener, Walter, 1951. P. A. Rydberg's "Report of a botanical exploration, made in western Nebraska in the summer of 1891, as a special agent of the [U. S.] Department of Agriculture." *Nebraska History* 32: 217–239, 306–331

Lewis, Walter. H., Barbara Ertter, A. Bruneau. 2014. *Rosa,* in *Flora of North America* 9: 75–119

Locklear, James H., 2013. Taxonomic identity and historical accounts of *Dalea cylindriceps* (Fabaceae), a species of conservation concern in the Great Plains (U.S.A.). *Journal of the Botanical Research Institute of Texas* 7(2): 879–890.

Perkey, Elton A., 1995. *Perkey's Nebraska Place Names* (revised ed.). Nebraska State Historical Society, 226 pp.

Porter, Thomas C., and John M. Coulter, 1874. Synopsis of the Flora of Colorado. *United States Geological and Geographical Survey of the Territories,* Miscellaneous Publications 4. Government Printing Office, Washington, D.C. 179 pp.

Prothero, Donald R. & Robert J. Emry. 2004. The Chadronian, Orellan, and Whitneyan North American land mammal ages, pp. 156-168 *in* M. O.Woodburne, ed., *Late Cretaceous and Cenozoic Mammals of North America.* Columbia University Press, 391 pp.

Rydberg, Per Axel. 1895a. Report of a Botanical Exploration of the Sand-hill Region of Central Nebraska made in the summer of 1893. M.A. thesis, University of Nebraska, June 1895.

Lincoln, 221 pp. http://digitalcommons.unl.edu/bioscidiss/83

Rydberg, Per Axel. 1895b. Flora of Nebraska, part 21: Rosales. *University of Nebraska Botanical Seminar* 26: 1–82.

Rydberg, Per Axel. 1895c. Flora of the Sandhills of Nebraska. *Contributions of the United States National Herbarium, Washington* 3: 133–203

Rydberg, Per Axel. 1900. Studies in the Rocky Mountain Flora 3. *Bulletin of the Torrey Botanical Club* 27: 614–636.

Rydberg, Per Axel. 1907. Scandinavians who have contributed to the knowledge of the flora of North America. *Augustana College* [Illinois] *Library Publications* 6: 41–43.

Rydberg, Per Axel. 1932. *Flora of the Prairies and Plains of Central North America.* New York Botanical Garden. 969 pp.

Tiehm, Arnold, and Frans A. Stafleu. 1990. Per Axel Rydberg: A Biography, Bibliography and List of His taxa. *Memoirs of the New York Botanical Garden* 58. 75 pp.

Urbanek, May. 1988. *Wyoming Place Names.* Mountain Press Publishing Co., Missoula, Montana. 233 pp.

Webber, Herbert J. 1890. Catalogue of the Flora of Nebraska. *Nebraska State Board of Agriculture, Report for 1889*: 36–162.

Winona [Minnesota] *Republican-Herald,* June 1, 1929. 3-year illness ends in death of J. M. Holzinger.

Wyoming State Library. 2016. Wyoming places: Sixtysix Mountain. http://wyomingplaces.pbworks.com/w/page/12716340/Sixtysix%20Mountain

www.ingramcontent.com/pod-product-compliance
Lightning Source LLC
LaVergne TN
LVHW070954210926
848482LV00002B/210